T0202090

The Birth of an Indian Profession

The Birth of an Indian Profession

Engineers, Industry, and the State, 1900–47

APARAJITH RAMNATH

OXFORD
UNIVERSITY PRESS

OXFORD
UNIVERSITY PRESS

Oxford University Press is a department of the University of Oxford.
It furthers the University's objective of excellence in research, scholarship,
and education by publishing worldwide. Oxford is a registered trademark of
Oxford University Press in the UK and in certain other countries.

Published in India by
Oxford University Press
2/11 Ground Floor, Ansari Road, Daryaganj, New Delhi 110 002, India

ISBN-13: 978-0-19-946987-1
ISBN-10: 0-19-946987-3

Typeset in ScalaPro 10/13
by The Graphics Solution, New Delhi 110 092
Printed in India by Rakmo Press, New Delhi 110 020

To the memory of
Ramchander
Jyothishmathi and Nalinikanta Rao
Visalakshi and Padmanaban

CONTENTS

FIGURES

TABLES

ACKNOWLEDGEMENTS

This book, the roots of which lie in my PhD research, could not have been written without the support and guidance of my doctoral advisors at Imperial College, David Edgerton and Abigail Woods. They have been equally generous with advice on this book in the years after I ceased formally to be their student. I am also indebted to Sabine Clarke and Sloan Mahone at Oxford, who supported my wish to go on to doctoral work, and to Meenakshi Raman and Sangeeta Sharma at the Birla Institute of Technology and Science, Pilani, who did likewise when I first contemplated postgraduate study.

Several historians have taken time out to give useful advice and encouragement. I particularly wish to thank Ross Bassett, Pratik Chakrabarti, Ralph Desmarais, Deepak Kumar, John Bosco Lourdusamy, Yogesh Mishra, Jahnavi Phalkey, Dhruv Raina, and Waqar Zaidi. I am grateful to Ramachandra Guha for talking to me, an unknown reader who grew up on his columns, over lunch one day at the British Library, and for his subsequent advice on various matters. Colin Divall, Mark Harrison, Felicity Mellor, Andrew Mendelsohn, and Andrew Warwick, as examiners at different stages of my PhD, provided feedback that has been valuable in working on this book. Margret Frenz has been a friend and mentor for ten years now.

A book that is several years in the making requires substantial funding, and various entities have provided the necessary support. The Bharat Petroleum Scholarship, the Hans Rausing Scholarship, and a Goodenough College Bursary supported my postgraduate research. As I developed the manuscript after moving back to India, visits to the UK for further archival work were funded by the Charles Wallace India Trust and the French National Research Agency (under its project 'Engineers and Society in Colonial and Postcolonial India' [ENGIND]). Thanks to Richard Alford, Solomon Jayasingh, Vanessa Caru, and Naziha Attia for coordinating these grants. Finally, the Indian Institute of Management (IIM) Kozhikode provided funding to attend conferences around the world where I was able to present parts of this book and related research.

For guiding me through the archives and making available the materials that form the basis of this study, although not all of it has made it into the book, I am grateful to Antonia Moon, John O' Brien, and the tireless staff at the British Library, London; R. P. Narla and Deepti Sasidharan at the Tata Central Archives, Pune; Swarup Sengupta at the Tata Steel Archives, Jamshedpur; Ishwara Bhatt at BITS Pilani; R. P. Chatterjee at the Institution of Engineers (India), Kolkata; Anne Locker and others at the Institution of Engineering and Technology, London; Carol Morgan at the Institution of Civil Engineers, London; Adrian Clement and others at the Institution of Mechanical Engineers, London; Kevin Greenbank at the archives of the Centre of South Asian Studies, Cambridge; the staff at the Bodleian and India Institute Libraries at Oxford and the SOAS and Imperial College/ Science Museum libraries in London.

The institutions of St Cross College, Oxford, and Goodenough College, London—and the many good friends I made in each place— provided a sense of community during the earlier stages of my research, while IIM Kozhikode has been a model employer in giving me the latitude to work on my manuscript. The friendship and guidance of my fellow faculty members at the IIM have helped me settle into an academic career. I would like particularly to thank A.F. Mathew, who never tired of reminding me to work on my book; Anupam Das, who has been a great source of moral support; and the directors and deans of the Institute in the years I have worked there.

Through comments, conversations, or opportunities to present my work at various stages, many people have enriched the process of developing this book. They include Charu Singh and Mahesh Rangarajan at the Nehru Memorial Museum and Library, New Delhi; Ross Bassett at North Carolina State University and at the Society for the History of Technology's recent Annual Meetings in Albuquerque and

Singapore; David Sicilia and Christina Lubinski at the World Business History Conference in Frankfurt, 2014; John Lourdusamy at IIT Madras; Bhaskar Chakrabarti at IIM Calcutta; Arvind at IISER Mohali; Nitish Nair at the Shell Technology Centre, Bangalore; Hrishikesh Chennakesavula at Ingersoll-Rand, Bangalore; Ines Zupanov, Manju Ludwig and Roger Jeffery at the 4th European South Asia PhD workshop in Heidelberg.

Joel Cabalion kindly introduced me to Roland Lardinois, from whom Vanessa Caru took up the baton to head up a project on Indian engineers, funded by the French National Research Agency. It has been a pleasure to get to know Roland and other members of the project, while I have learnt a great deal, especially about the public works engineers of Bombay Presidency, from conversations with Vanessa.

Several friends acted as foster family during my years in (and subsequent visits to) the UK. I would have been adrift without Zhiyu Chen, Xiaoyang You, Rukshan Batuwita, Shashika Subasinghe, Debapratim De, and Naina Bhattacharya. In India, friends and family in various cities put me up on my research visits, took a keen interest in my work, saved books and newspaper articles for me, and participated in many long conversations. They include R. V. S. Mani in Chennai; P. Balakrishna and Rashmi Balakrishna in Delhi; Hrishikesh Chennakesavula in Jamshedpur and Bangalore; Arnab Chakraborty in Kolkata; and R. Kumar and Usha Kumar in Pune. Anand Bhaskaran's company was invaluable during an otherwise secluded year spent writing in Chennai, as was Smriti Naropanth's encouragement many years ago, when book-writing was but a gleam in my eye.

Special thanks to the team at Oxford University Press (OUP), India, the publishers of this book, for bringing the project to fruition. In revising the manuscript, I have also benefited from the incisive comments received from the two external referees selected by OUP.

My parents and my sister Shreya have been invested in this project from the start. Without their patience, even indulgence, and the encouragement of my extended family, I would have found writing this book—not to speak of my somewhat reckless change of career path from engineering to history—far more arduous. Akshaya has brought more warmth and laughter to my life, and shown me that one can be a conscientious scholar without being pompous.

Countless other people have helped in many ways, but lest these acknowledgements turn into a boast for the amount of work I have done, I will content myself with thanking them in person. Naturally, any shortcomings that remain are my responsibility alone.

Introduction

Engineers in India, 1900–47

In 1925, the Calcutta-based weekly *Indian Engineering* published a letter from a reader identified simply as 'C.E.' 'Of the numerous public
bodies in this country holding important and responsible positions,'
the letter began, 'there are few so characteristically modest as the
members of the engineering profession, both European and Indian.'
These engineers toiled selflessly, receiving their due neither from the
state nor from 'a certain class of extremist politicians whose sole aim
is destruction, not construction'. The letter continued:

> Where would India be now were it not for its railways and gigantic irriga-
> tion works?... What would the various industries which are in existence
> have done without its engineers; they are the backbone of them. All indus-
> trial and commercial enterprises are indebted to them. The engineering
> profession is one of the fundamental professions in the world. India cannot
> have too many engineers, there is work for all who will patiently study the
> many subjects connected with engineering, and set to work in right earnest
> to carve out a brilliant career for themselves.[1]

As essential components of the Public Works Department (PWD)
and the railways, engineers were at the heart of Indian government
bureaucracies in the period between 1900 and 1947. They also played

---

[1] Letter to the Editor from 'C.E.', *Indian Engineering*, 7 November 1925, p. 260.

a central role in the newly emerging large-scale industries. Theirs was a profession in the throes of a significant transformation, and the letter from 'C.E.' (no doubt an interested party) hints at the anxieties that accompanied it. In the interwar period, engineers working in India were increasingly conscious of a collective identity; a rising proportion of 'native' Indian engineers were working alongside British expatriates; and government and industrial engineers alike operated in the midst of continuous political change as the colonial era entered its twilight. Despite their importance, however, there has been no systematic study of the engineering profession in these eventful years.[2]

This book is the first extensive history of engineers working in modern India, charting the development of the profession from 1900 to 1947. Including in its scope engineers in the public works, railways, and private industry, this volume examines how changes in the engineering profession[3] in this period were caused by and contributed to two important transformations in Indian society. The first of these was industrialization, or the growth of private, especially large-scale, industry in the interwar years. The second was Indianization, a term that refers simultaneously to the increasing proportion of 'native' Indians in government services and private firms, and to the political ideal this process represented, as one of the central demands that nationalist leaders made of the colonial government. This process brought into focus ideas of race and technical expertise, and revealed the particular areas where the colonial government most wanted to maintain British engineers in charge. Situating the experience of an extremely important, yet under-studied, group of technical practitioners within the political and economic history of India, this study of engineers contributes to the historiography of science and technology in India.

The history of science and technology in India has formed a distinct field of enquiry since at least the 1990s. The pioneering works in this field, which mostly addressed the long nineteenth century,[4]

---

[2] There have, however, been works dealing partially with engineers, or with particular types of engineers, as discussed later in this chapter.

[3] Later in this chapter I discuss the sense in which I use the term 'profession' in this book.

[4] David Arnold wrote in 2000 that 'the science, technology and medicine of the period between the outbreak of the First World War and Indian independence have,

examined the relationship between science and colonialism. Investigating the nature of science in a colonial location and the manner of its growth, they argued that science was not merely a European entity transplanted into Indian soil. Instead, Indians 'respon[ded]' to this science;[5] it had to be 'translated' for the colony and 'staged' in a way that challenged its purported unemotional and objective nature;[6] and there were varied forms 'of exchange between modern science and so-called traditional knowledge forms'.[7] Ultimately, though, these works stressed the notions of limitation, dependence, and subordination to the imperial metropolis.[8] Where it addressed the twentieth century, the literature demonstrated the centrality of science and technology in nationalist intellectuals' discourse on the creation of an Indian 'modernity',[9] despite their varying (and sometimes internally

---

as yet, attracted little scholarship'. David Arnold, *Science, Technology and Medicine in Colonial India*, The New Cambridge History of India, III.5 (Cambridge: Cambridge University Press, 2000), p. 225. I use Hobsbawm's term 'long nineteenth century' loosely to mean the nineteenth century and the pre–Great War years of the twentieth century.

[5] Deepak Kumar, *Science and the Raj, 1857–1905* (Bombay and Oxford: Oxford University Press, 1995), chapter 6: 'Response and Resistance'.

[6] Gyan Prakash, *Another Reason: Science and the Imagination of Modern India* (Princeton: Princeton University Press, 1999), esp. chapter 2.

[7] S. Irfan Habib and Dhruv Raina (eds), *Social History of Science in Colonial India* (New Delhi: Oxford University Press, 2007), 'Introduction', p. xxiii.

[8] Deepak Kumar's *Science and the Raj* shows that colonial officials at various levels in the bureaucracy had their own ideas about the administration of science, although they were frequently frustrated by the rigid hierarchies of colonial India. The frustrations and lack of autonomy of science in colonial India also form a recurring theme in Pratik Chakrabarti's *Western Science in Modern India*. Similarly, several essays in Roy MacLeod and Deepak Kumar's edited volume *Technology and the Raj* are studies of the limited scope of the science and technology project in India, which was always subject to the needs of the colonial government. Pratik Chakrabarti, *Western Science in Modern India: Metropolitan Methods, Colonial Practices* (Delhi: Permanent Black, 2004); Roy MacLeod and Deepak Kumar (eds), *Technology and the Raj: Western Technology and Technical Transfers to India 1700–1947* (New Delhi, Thousand Oaks, and London: Sage Publications, 1995). See, for example, the essays by Arun Kumar (pp. 216–32) and V. V. Krishna (pp. 289–323) in *Technology and the Raj*.

[9] In Gyan Prakash's account, the modern Indian state was conceptualized and configured within the boundaries of the action of 'technologies of government' (Gyan Prakash, *Another Reason*, chapter 6). Pratik Chakrabarti writes that '[i]ndustrialism provided the ultimate confirmation of the acceptance and habitation of science in

contradictory) views on its role. Thus, physicist M. N. Saha was strongly in favour of science-based industry; Mahatma Gandhi did not support industrialization except under certain conditions; chemist P. C. Ray, who established an important industrial enterprise, was nevertheless a proponent of the Gandhian *charkha*; and geologist P. N. Bose viewed industrialization as, at best, a necessary evil.[10]

While these works form an indispensable starting point for the study of science and technology in India, the focus on particular aspects of questions of colonialism, nationalism, and modernity has led to certain limitations in the historiography. First, there has been a comparative neglect of the study of pre-Independence institutions, practices, and professional groups in favour of the analysis of the discourses of colonial elites and nationalist intellectuals.[11] The individuals studied, moreover, are a small group of *political* and *scientific* intellectuals like Gandhi, Saha, and Ray as opposed to, for instance, the much larger community of professional engineers. Second, the reliance on the analytical frameworks of 'colonialism' and 'nationalism' in interpreting science, technology, and medicine in India has resulted in a starkly dichotomous picture: a nineteenth century

India' (Chakrabarti, *Western Science in Modern India*, p. 297). The literature's focus on science in relation to 'modernity' is evident in the titles of books on science in colonial India: for instance, Prakash, *Another Reason: Science and the Imagination of Modern India*; Chakrabarti, *Western Science in Modern India*; Dhruv Raina, *Images and Contexts: The Historiography of Science and Modernity in India* (Delhi and Oxford: Oxford University Press, 2003).

[10] Deepak Kumar, 'Reconstructing India: Disunity in the Science and Technology for Development Discourse, 1900–1947', *Osiris*, 2nd series, vol. 15 (2000): 241–57; Chakrabarti, *Western Science in Modern India*, chapter 8.

[11] By contrast, a recent study of the post-Independence period by Amit Prasad has considerable detail on scientists' careers, professional networks, and concerns with obtaining funding. However, it is intended as a 'transnational' story, not an India-specific one. Amit Prasad, *Imperial Technoscience: Transnational Histories of MRI in the United States, Britain, and India* (Cambridge, MA: MIT Press, 2014, Kindle edition). See also Jahnavi Phalkey's book exploring the institutionalization of nuclear physics in India from the 1930s to the post-Independence period, through the stories of three prominent scientists (M. N. Saha, C. V. Raman, and H. J. Bhabha) and their attempts to procure particle accelerators for their respective research centres. The building of research groups and individual careers is as important to this narrative as its protagonists' concerns with modernity or state-building. Jahnavi Phalkey, *Atomic State: Big Science in Twentieth-Century India* (Ranikhet: Permanent Black, 2013).

characterized by 'tools of imperialism'[12] (and mostly European actors) and a mid-twentieth century by tools of development and nationhood (and mainly Indian actors).[13] The consequence is that we know relatively little about the dynamic and changing role of the colonial state in science and technology for nearly half of the twentieth century.

But the story of knowledge and expertise in India must go beyond describing its relationship with colonialism and highlighting its circumscribed and dependent nature. It must address other questions as well: How did elite scientists and engineers engage with the opportunities that existed within a system that was not entirely autonomous?[14] How did a mixed engineering profession (comprising Europeans and Indians) develop in late colonial India? The story needs to dwell less on how limited developments in Indian science and technology were in comparison to some idealized trajectory, and more on its actual development.

This book is about a prominent group of scientific/technical practitioners, their professional concerns, and their work culture. Reconstructing engineers' professional lives using their memoirs, government records, official reports, institutional and industrial archives, and trade journals, this book asks the following questions: How were engineers' careers related to the political and economic changes occurring in interwar India? How did this relate to the ideological changes in this period—to what extent were engineers concerned with justifying imperialism or promoting nationalism? What was distinctive about the Indianization of engineering in government services as opposed to private industry?

---

[12] To quote Daniel R. Headrick's phrase. See his 'The Tools of Imperialism: Technology and the Expansion of European Colonial Empires in the Nineteenth Century', *Journal of Modern History*, vol. 51, no. 2 (June 1979): 231–63.

[13] For convenience, I follow in this book the convention of referring to the two major races represented in British/colonial India by the terms 'Indian' and 'European'. Likewise, I use the term 'race' as a shorthand for the racial label applied to an individual for administrative purposes during the colonial period.

[14] Aparajith Ramnath, 'Engineers in India: Industrialisation, Indianisation and the State, 1900–47' (PhD thesis, Imperial College London, 2012), chapter 1. See also Jahnavi Phalkey, 'Focus: Science, History, and Modern India. Introduction', *Isis*, vol. 104, no. 2 (June 2013): 330–6. Ramnath, 'Engineers in India', chapter 1 also contains a detailed analysis of the contributions and limitations of the literature.

The broad narrative is as follows. At the turn of the twentieth century, most engineers working in India were government employees, either in the PWD or the railways. These organizations grouped their engineers—who were predominantly British expatriates—into hierarchical bureaucratic services. Engineering education in India was limited to a few government-run colleges and focused on civil engineering.

Four decades later, as I demonstrate, the picture had changed considerably. The most striking development was the Indianization of the profession: a substantial proportion of British engineers had been replaced by 'native' Indians, who now made up well over 50 per cent of the elite government engineering services. Simultaneously, a nascent Indian identity emerged among engineers, who had previously seen themselves as belonging to a larger, empire-wide profession centred in London. The second fundamental change was related to industrialization. The profession in this period diversified to include a larger proportion of industrial engineers, even as education broadened from a focus on civil engineering (and the inculcation of gentlemanly qualities) to include emerging branches such as electrical engineering, mechanical engineering, and metallurgy (while emphasizing physical toil and economic productivity).

This transformation was neither automatic nor uneventful. I argue that it was the result of a variety of factors, including vociferous political agitation, important changes in the structure of the colonial state, the effect of two World Wars, the changing economic relationship between Britain and India, and the resourcefulness of Indians who went beyond the British Empire to study industrial engineering in countries like Germany and the United States. Official commissions were set up to look into the question of Indianization; legislators in India and Britain debated the issue keenly; and constitutional reforms caused the reorganization of government bureaucracies and a fresh articulation of the roles of engineers, revealing in the process contemporary ideas on race and technical expertise. If conventional wisdom holds that the Nehruvian era was crucial to the development of engineering in India,[15] this book argues that formative influences on the profession go further back, to the years between the World Wars.

[15] For a more nuanced version of this argument, see Dinesh C. Sharma, *Nehru: The Unlikely Hero of India's Information Technology Revolution*, NMML Occasional Paper, Perspectives in Indian Development, New Series 8 (New Delhi: Nehru Memorial

Two points are worth emphasizing in this narrative. First, the experience of technical practitioners in the period under study was more complex than what the existing historiography, with its emphasis on *scientist*-intellectuals and the discourse on modernity and nation, would suggest. By studying both British and Indian practitioners in parallel and in relation to each other—an approach that has rarely been adopted before[16]—I demonstrate that British engineers, despite Indianization, continued to play an important role, especially in defining the culture of the engineering services; and that the Indian engineers who rose to prominence were not necessarily political radicals or participants in the extensive debates on science and modernity. Second, the book shows, through its detailed study of the dynamics of the Indianization process, that the interwar experience of engineers cannot be understood adequately without taking into account the heterogeneous nature of the colonial state. The interwar India in which both British and Indian engineers worked cannot be captured solely by notions of colonialism or nationalism, but needs to be understood

---

Museum and Library, 2013). Sharma argues that the roots of contemporary India's 'information technology revolution' go back to the Nehruvian era and the scientific/technological institutes (especially the Indian Institutes of Technology) nurtured in those years, which enabled the development of computer science and engineering in India.

[16] Pratik Chakrabarti, in his *Western Science in Modern India*, argues for the advantage of studying the connections between 'the practice of science by Europeans and by Indians' (p. 24). While his book studies both European and Indian practitioners of science, it is the Indians who take centre stage in the latter chapters covering the twentieth century. Studies in which engineers play an important part tend (as a result of the periods they study and/or the questions they explore) to be confined either to British or Indian engineers. The former feature prominently in David Gilmartin, 'Scientific Empire and Imperial Science: Colonialism and Irrigation Technology in the Indus Basin', *The Journal of Asian Studies*, vol. 53, no. 4 (November 1994): 1127–49; two Indian engineers are the protagonists in Daniel Klingensmith's analysis of 'nationalist engineering' in late-colonial and Independent India—see chapter 5 of his *'One Valley and a Thousand': Dams, Nationalism, and Development* (New Delhi: Oxford University Press, 2007). For a rare exception, see Y. Srinivasa Rao, 'Electrification of Madras Presidency, 1900–1947' (PhD thesis, Indian Institute of Technology Madras, 2007). One of Rao's arguments is that British and foreign-trained Indian engineers tended to see electricity as ideal for large-scale industrialization, while locally trained Indian engineers believed in electrification of the rural areas with a view to promoting agriculture, cottage industries, and self-sufficient villages.

also in terms of the continuously negotiated and redefined conception of a self-governing dominion within the British Empire, a conception that had currency for much of this period.[17]

In the following section I elaborate upon this approach to the study of engineers in India, and discuss a small number of works upon whose contributions it builds.

HISTORY OF SCIENCE AND TECHNOLOGY AS INDIAN HISTORY

David Edgerton has argued recently that histories of technology must not just be studies of technology (in which history provides the examples), but of 'technology *in history*—asking questions about the place of technology in wider historical processes'.[18] In the field as it stands, however, 'historical questions are often secondary', while the main object of study is something called 'the question of technology', which refers to 'the nature of technology and its relations to wider culture'.[19] Taking an uncritical view of history as providing the 'context' for a particular technology obscures the fact that the very context should be an object of analysis. Writing 'post-contextual' histories of technology, then, can advance our understanding of history itself.[20]

My approach in this book is in keeping with these observations. I seek to place the history of science and technology in India firmly within Indian history. I am concerned not only with 'colonial technology' or 'Indian engineering', but also with what we can learn, by studying an important group of practitioners, about the nature of the Indian economy, polity, and society in the interwar period. The colonial state was not a homogeneous entity placed in direct opposition to

---

[17] See R. J. Moore, 'The Problem of Freedom with Unity: London's India Policy, 1917–47', in *Congress and the Raj: Facets of the Indian Struggle 1917–47*, edited by D. A. Low, 2nd edn, pp. 375–403 (New Delhi: Oxford University Press, 2004), for a discussion of the dominion idea.

[18] David Edgerton, *The Shock of the Old: Technology and Global History since 1900* (London: Profile, 2008 [2006]), p. 211. Emphasis mine.

[19] David Edgerton, 'Innovation, Technology or History: What Is the Historiography of Technology About?', *Technology and Culture*, vol. 51, no. 3 (July 2010): 680–97, here p. 691. Edgerton quotes Thomas Misa as using the phrase 'the question of technology', following Martin Heidegger.

[20] Edgerton, 'Innovation, Technology or History', pp. 694–6.

nationalist politicians. Constitutional reform and the stepwise introduction of provincial autonomy gave elected Indian ministers a measure of executive power in the provinces from 1919; Indians were also
able to join the Central Legislative Assembly in large numbers, where
they could act as a check on the executive, composed mainly of British
officials.[21] Studying engineers in various sectors not only illuminates
how the evolving colonial state and its economic policies shaped the
engineering profession, but also sheds light on the workings of the
state itself in this transitional period of Indian history.

In adopting this approach, I aim to extend the fruits of a small
number of works on the interwar period, which illustrate how the
changing colonial state and the institutional basis of science and technology were closely linked. David Arnold's *Science, Technology and
Medicine in Colonial India* is exceptional in noting that '[a] history of
science in India must also be a history of India, not merely a history
of the projection of Western science onto India'[22]—an injunction it
follows with considerable success. Arnold demonstrates the importance of the growing presence of Indians in the legislative machinery
in the twentieth century. Agitations by Indian members of the provincial legislatures for the teaching of local systems of medicine led
to the creation of a School of Indian Medicine in Madras Presidency
in 1924; increasing provincial autonomy also created a tug of war
between the proponents of centralized and decentralized scientific
research, leading to a mix of all-India scientific services and central
institutes on the one hand and provincial institutes/university departments on the other.[23] A recent paper by Pratik Chakrabarti argues that
the provincialization of medical research and the Indianization of the
Indian Medical Service (IMS) in the interwar years prompted British
medical officers increasingly to seek refuge in the idea of centralized
institutes, of which they would retain control. Indian politicians and
university professors thought differently, and successfully opposed a
proposal to set up a Central Medical Research Institute in the hill

---

[21] See, for instance, Burton Stein, *A History of India*, 2nd edn, edited by David
Arnold (Oxford: Wiley-Blackwell, 2010), chapter 7, for an overview of the constitutional
reforms in this period. The changing structure of government is discussed in the first
chapter of this volume.

[22] Arnold, *Science, Technology and Medicine in Colonial India*, p. 2.

[23] Arnold, *Science, Technology and Medicine in Colonial India*, chapter 6.

station of Dehra Dun.[24] The sociologist Roger Jeffery's early work on the medical profession in India is important in this context. In two papers published in the late 1970s, Jeffery studied rivalries within and without the profession of 'allopathic' (that is, 'Western', but not homoeopathic) doctors in twentieth-century India. Jeffery shows that the course of Indianization of the medical profession in the interwar years and the subsequent 'deprofessionalisation' of allopathic doctors in the decades after Independence were closely tied to changes in the nature of the state (the colonial state, with constitutional reforms, before Independence and the Indian republic thereafter).[25]

As a corollary to the increasingly heterogeneous state, I take the position in this book that not all science and technology in the interwar years is best understood as 'colonial' (or, for that matter, as putatively 'national'). As Mark Harrison has argued, the intimate relationship of science with colonialism is not to be denied, but we must note that the 'nature [of science] was not defined by colonialism alone'.[26] Indeed, recent works have emphasized the importance of international, extra-imperial networks in the growth of science and technology in India. Ross Bassett, for instance, studies a group of Indians who studied at the Massachusetts Institute of Technology before Independence, funded largely by princely rulers. These engineers, some of whom set up Swadeshi industries or joined the nationalist movement upon their return to India, 'serve as a significant point of origin for a tech-nological identity defined in relation to the United States'.[27] The fifth chapter of this book, which focuses on the technical experts of the Tatas' steel company, explores another instance of this phenomenon

[24] Pratik Chakrabarti, '"Signs of the Times": Medicine and Nationhood in British India', *Osiris*, vol. 24, no. 1 (2009): 188–211.

[25] Roger Jeffery, 'Recognizing India's Doctors: The Institutionalization of Medical Dependency, 1918–39', *Modern Asian Studies*, vol. 13, no. 2 (1979): 301–26; Roger Jeffery, 'Allopathic Medicine in India: A Case of Deprofessionalisation?', *Economic and Political Weekly*, vol. 13, no. 3 (21 January 1978): 101–3, 105–13.

[26] Mark Harrison, 'Science and the British Empire', *Isis*, vol. 96, no. 1 (March 2005): 56–63, here p. 63.

[27] Ross Bassett, 'MIT-Trained Swadeshis: MIT and Indian Nationalism, 1880–1947', *Osiris*, vol. 24, no. 1 (2009): 212–30; quoted text from p. 214. Bassett has recently brought out a book on Indian graduates of MIT. Ross Bassett, *The Technological Indian* (Cambridge, MA, and London: Harvard University Press, 2016, Kindle edition).

in the shape of the American and America-trained Indian engineers who were central to the running of the works.

This book is also concerned with another aspect of technology in interwar India—the role of race. Ideas associating certain races with scientific and technical ineptitude abounded in the nineteenth and twentieth centuries. Black Americans were considered inherently less inventive than their white compatriots, were not allowed to fly military aircraft in the interwar period, and were not employed as telephone operators at Bell until after World War II. Significantly for colonial science and technology, when 'Western' technologies were 'transferred' to other, usually colonial, locations, Europeans went along to operate them, as in the case of British and French pilots on the ships sailing through the Suez Canal. As of 1917–18, all railway officers in the Dutch East Indies were Europeans.[28] A few scholars have studied similar ideas operating in colonial India. They have argued, for instance, that prevailing views of Indians as lacking in technical aptitude were related to the European domination of the upper railway ranks in pre–Great War India and the general reluctance of the colonial government to set up technical education facilities for Indians.[29] Witnesses called by the Slacke Committee (1912), which had been set up to provide advice on the proposal for a technical institution in Calcutta, 'held that Bengalis as a race were unfit for practical work. It was because they lacked interest and stamina that they could not find employment'.[30] Racial prejudice in one form or another also meant that few Indians held responsible positions in the government's scientific and medical

[28] Edgerton, *Shock of the Old*, pp. 131–6.

[29] Daniel R. Headrick, *The Tentacles of Progress: Technology Transfer in the Age of Imperialism, 1850–1940* (New York and Oxford: Oxford University Press, 1988), chapter 9.

[30] Aparna Basu, 'Technical Education in India, 1854–1921', in *Essays in the History of Indian Education* (New Delhi: Concept, 1982), pp. 39–59, here p. 51. On the Slacke Committee, see also Basu, 'Technical Education in India', p. 48 and endnote 44 on p. 58. This characterization of Bengalis had a long lineage, shaped, as David Arnold argues, by the region's frequent experience of malaria, which was taken to have made Bengalis weak and unmanly. Arnold refers to both European and Bengali perceptions of Bengali racial identity. David Arnold, '"An Ancient Race Outworn": Malaria and Race in Colonial India, 1860–1930' in *Race, Science and Medicine, 1700–1960*, edited by Waltraud Ernst and Bernard Harris (London and New York: Routledge, 1999), pp. 123–43.

services until at least 1920.[31] David Arnold's recent study of 'everyday technology' shows that stereotypes were not restricted to the official mind. Sales strategies and representations in marketing material revealed an assumption that Europeans and Eurasians were more suitable and likely users of technologies such as sewing machines and typewriters. Further, ideas of class and gender operated in concert with beliefs regarding race, and, crucially, the 'close identification of machines with gender and ethnicity was not a colonial invention alone but drew, too, upon notions of gender and community long established in Indian society'.[32]

While many of these accounts deal predominantly with the long nineteenth century, this book examines the issue of race with a particular emphasis on interwar India. Prejudices about Indians continued in this period but took on a slightly different character in relation to engineers, as I discuss in the third and fourth chapters. My focus on Indianization and industrialization demonstrates how the engineering profession shaped and was shaped by important political developments. In doing this I not only pay attention to the differing experiences of practitioners depending on their race, but also study engineers as an integral part of wider Indian history. In the next section I introduce the concepts of and literature on Indianization and industrialization in pre-Independence India.

INDIANIZATION AND INDUSTRIALIZATION

Indianization, or the increasing of the number of 'native' employees in government service, was one of the central issues in Indian politics from 1858 (when Crown Rule began) to 1947 (when Independence was achieved).[33] Despite its political importance, not many historians have studied the question in detail. A few studies concerned indirectly with the issue of Indianization appeared in the 1970s and 1980s. These were focused on the Indian Civil Service (ICS) and the

[31] Deepak Kumar, 'Racial Discrimination and Science in Nineteenth-Century India', *Indian Economic and Social History Review*, vol. 19 (1982): 63–82; Arnold, *Science, Technology and Medicine in Colonial India*, chapter 5.

[32] David Arnold, *Everyday Technology: Machines and the Making of India's Modernity* (Chicago and London: University of Chicago Press, 2013), chapter 3. Quoted text on p. 94.

[33] For a brief note on the provenance of the term 'Indianization', see Chapter 1.

changes that occurred in its composition and culture in the final decades of British rule. David Potter argued that one of the important reasons for the withdrawal of Britain from India in 1947 was that the British personnel it relied on to staff the civil services were dwindling, as fewer Britons opted to apply for Indian service from 1919 onwards. Asking why the popularity of the ICS fell among British applicants, he suggested that it was the inadequate terms of employment that worried them more than constitutional changes in India and the rise of the nationalist movement.[34] Ann Ewing, in contrast, showed convincingly that political uncertainty, the hostility of nationalist politicians towards the elite European-dominated nature of the ICS, and the decrease in powers of ICS officers under the changing system of government in the interwar years were a serious source of 'discontent' in the ICS and of worry to the government in Britain and in India.[35] A more recent book by Malti Sharma tracing debates on the Indianization of the ICS from 1858 to 1935 demonstrates that the expanding role of Indians in the Indian legislative councils in the twentieth century provided a key impetus to Indianization in the ICS.[36]

The ICS may have been the most prestigious and powerful, but there was a multiplicity of government services, and Indianization as a political issue applied to all of them. A rare example of a study that considers multiple services is J. D. Shukla's 1982 book on Indianization. Shukla analyses Indianization across several 'All-India Services'[37] such as the Indian Police, Indian Service of Engineers (ISE),

---

[34] David C. Potter, 'Manpower Shortage and the End of Colonialism: The Case of the Indian Civil Service', *Modern Asian Studies*, vol. 7, no. 1 (1973): 47–73, esp. p. 54. On ICS recruitment, see Potter's article and also T. H. Beaglehole, 'From Rulers to Servants: The I. C. S. and the British Demission of Power in India', *Modern Asian Studies*, vol. 11, no. 2 (1977): 237–55.

[35] Ann Ewing, 'The Indian Civil Service 1919–1924: Service Discontent and the Response in London and in Delhi', *Modern Asian Studies*, vol. 18, no. 1 (1984): 33–53.

[36] Malti Sharma, *Indianization of the Civil Services in British India (1858–1935)* (New Delhi: Manak Publications, 2001). There exists also some work on the Indianization of the armed forces in India. See Chandar S. Sundaram, '"Treated with Scant Attention": The Imperial Cadet Corps, Indian Nobles, and Anglo-Indian Policy, 1897–1917', *The Journal of Military History*, vol. 77, no. 1 (January 2013): 41–70. Dr Sundaram's monograph on the Indianization of the army is due to appear in the near future. I thank him for our email correspondence on this subject.

[37] The term is explained in Chapter 1.

and Indian Educational Service, in addition to the ICS. His book is a valuable overview of the terms and recommendations of various government commissions set up to look into the organization of the services (including their racial composition) at the time of constitutional reforms, and shows the general trends in the changing proportion of Indians in various services.[38] However, as the ISE is only one of several technical and non-technical services covered, Shukla's book does not tell us much about the particular features of engineers' roles and experiences at the time of Indianization. Nor does it deal with the important cases of railway engineers (railway officers did not form an 'All-India Service'—see Chapter 1) and technical personnel in private industry. In this book I explore the Indianization not only of the ISE—whose officers staffed the PWD—but also of technical officers in the Indian railways (privately run as well as government-run) and of the technical experts of a prominent heavy industrial enterprise (Chapters 3, 4 and 5).[39] Further, I consider the Indianization of engineers not only in relation to the government services, but as a larger issue, closely related to the educational backgrounds, identities, and work culture of engineers in the first half of the twentieth century.

The other major process addressed in this book is the growth of industries, especially after World War I. The nature of 'industrialization' in interwar India is a much-discussed topic in the literature on economic history. As Rajnarayan Chandavarkar has noted, scholars have mostly conceptualized industrialization as an idealized, teleological process, against which India's interwar industrialization appears as incomplete or limited.[40] Thus, R. K. Ray is concerned with why the 'private corporate sector' failed 'to transform the economy from a predominantly agricultural to a predominantly industrial one', while

[38] J. D. Shukla, *Indianisation of All-India Services and Its Impact on Administration* (New Delhi: Allied Publishers, 1982).

[39] Ian Kerr and Daniel Headrick briefly comment upon Indianization in the railways, while Headrick also gives a short account of Indianization in private industry. I engage with these studies in Chapters 4 and 5. The works referred to are Ian J. Kerr, *Engines of Change: The Railroads That Made India* (Westport, CT and London: Praeger, 2007), chapter 6; Headrick, *The Tentacles of Progress*, chapter 9 and pp. 371–4.

[40] Rajnarayan Chandavarkar, 'Industrialization in India before 1947: Conventional Approaches and Alternative Perspectives', *Modern Asian Studies*, vol. 19, no. 3 (Special Issue, 1985): 623–68, esp. p. 664.

Dietmar Rothermund writes of 'the limits of industrialisation under colonial rule' in the long nineteenth century and of 'the Indian economy, which continued to suffer from the chronic disease of colonial paralysis' in the period after World War I.[41] The result, Chandavarkar argues, is an exercise in apportioning the blame for a supposedly incomplete industrialization 'between the baneful effects of colonial rule [and] the timeless torpor of Indian society'.[42] In the former category we might place the colonial government's fiscal policy and inadequate protection of Indian industry; in the latter low domestic demand, insufficient capital, and the reluctance of industrialists to take risks.[43]

Yet, if we were to ask a positive historical question (To what extent did industry grow?) instead of the negative one privileged by the literature (Why was industrialization limited/incomplete?), the findings of this literature could be seen in a different light, as providing considerable evidence for a marked growth in industry. R. K. Ray writes that by the time of Independence, 'India had a larger industrial sector, with a stronger element of indigenous enterprise, than most underdeveloped countries of the world'.[44] Macroeconomic data support this verdict. Tirthankar Roy has shown that large-scale industry in particular expanded: employment in this sector grew from 0.88 million in 1911 to 1.57 million in 1921, staying at that figure in 1931. Large-scale industry's share of industrial income had been 15 per cent in 1900; in 1947 it was around 40 per cent.[45]

[41] Rajat K. Ray, *Industrialization in India: Growth and Conflict in the Private Corporate Sector 1914–47* (New Delhi: Oxford University Press, 1982 [1979]), p. 1; Dietmar Rothermund, *An Economic History of India: From Pre-Colonial Times to 1991*, 2nd edn (London: Routledge, 1993), quoted text from the title of chapter 5 and from p. 64.

[42] Chandavarkar, 'Industrialization in India before 1947', pp. 635, 664.

[43] A. K. Bagchi's argument on protection, as described in Tirthankar Roy, *The Economic History of India 1857–1947*, 2nd edn (New Delhi: Oxford University Press, 2006), p. 263; Rothermund, *An Economic History of India*, pp. 61–5; Ray, *Industrialization in India*, pp. 3, 81, chapter 4.

[44] R. K. Ray, *Industrialization in India*, p. 1.

[45] T. Roy, *The Economic History of India*, pp. 186–7 (including Table 6.3) and chapter 6 in general.

It is this aspect of India's industrialization that is relevant to my enquiry in this book. There may have been many failed ventures, and India may still have been a mainly agricultural economy at the time of Independence; nevertheless, industries, especially large-scale industries, grew—and with them the importance of industrial engineers. While historians have studied class identity among industrial workers,[46] the crucial role of these engineers and technical experts in the interwar growth of industries is largely neglected in the literature.[47] (Also from a class perspective, engineers, despite being the link between capital and labour in industries, and between political higher-ups and subordinate employees in key government services, have not featured prominently in historical studies of the Indian middle classes.[48]) The present book, through its detailed study of the role of technical experts in a prominent interwar industrial enterprise (the Tatas' steel company), highlights this important dimension of the industrialization process (Chapter 5). At the same time, it also pays attention to the effect of growing industry on engineering as a profession in interwar India (Chapter 2).

ENGINEERS IN INTERWAR INDIA

While we do not have a pan-Indian historical account of engineers in the interwar period, the few existing works that study engineers directly or indirectly show that a wide variety of engineers worked in pre-Independence India. Their roles, too, have been variously understood. Scholars have written studies of the engineering colleges at Roorkee, Sibpur, Poona, and Madras, established in the nineteenth

---

[46] Rajnarayan Chandavarkar, *The Origins of Industrial Capitalism in India: Business Strategies and the Working Classes in Bombay, 1900–1940* (New Delhi: Foundation Books [in arrangement with Cambridge University Press], 1994); Parimal Ghosh, *Colonialism, Class and a History of the Calcutta Jute Millhands 1880–1930* (Hyderabad: Orient Longman, 2000).

[47] A key exception is Ross Bassett's latest book, which studies American-trained Indian technologists over the nineteenth and twentieth centuries, many of whom returned to India and joined or attempted to set up industrial enterprises. Bassett, *The Technological Indian*.

[48] See, for instance, the eclectic collection of papers in *The Middle Class in Colonial India*, edited by Sanjay Joshi (New Delhi: Oxford University Press, 2010).

century by the colonial government[49] and charged primarily with supplying engineers and engineering subordinates to the PWD.[50] Ian Kerr's account of the construction of the Indian railways, in essence a labour history, is concerned with how the mostly European railway engineers handled the 'problems of management', relying, for instance, on Indian contractors to mobilize 'native' labour.[51] David Gilmartin's analysis of 'colonialism and irrigation technology' in the Indus River Basin shows engineers as key members of government bureaucracies. Gilmartin argues that after the construction—in the 1880s—of a network of irrigation canals in the Indus Basin, irrigation engineers and civil administrators' ideas of how to run the canal system clashed: while the former believed in engineering solutions, the latter believed in paying heed to local customs and ideas of community and hereditary rights.[52] While the histories just mentioned mainly study the nineteenth century, a valuable work dealing with a later period is Daniel Klingensmith's chapter on A. N. Khosla and Kanwar Sain as practitioners of 'nationalist engineering' in his book on dams and ideas of development in India and the USA. Here, Klingensmith is concerned primarily with the late- and post-colonial India of the 1940s and 1950s.[53] A rare exception in covering the interwar period, although it focuses on one province, is a doctoral thesis by Y. Srinivasa Rao on the introduction and development of electricity in the Madras Presidency from the year 1900 to 1947. Engineers, both British and Indian, along with provincial legislators, appear in Rao's

---

[49] See, for example, K. V. Mital, *History of the Thomason College of Engineering (1847–1949): On Which Is Founded the University of Roorkee* (Roorkee: University of Roorkee, 1986); John Bosco Lourdusamy, 'College of Engineering, Guindy, 1794–1947', in *Science and Modern India: An Institutional History, c. 1784–1947*, vol. 15, edited by Uma Das Gupta, Part 4 of Series: 'History of Science, Philosophy and Culture in Indian Civilization' (Delhi: Pearson Longman, 2011), chapter 15 (pp. 429–50).

[50] Arun Kumar, 'Colonial Requirements and Engineering Education: The Public Works Department, 1847–1947' in *Technology and the Raj: Western Technology and Technical Transfers to India 1700–1947*, edited by Roy MacLeod and Deepak Kumar (New Delhi, Thousand Oaks, and London: Sage Publications, 1995), pp. 216–32.

[51] Ian J. Kerr, *Building the Railways of the Raj, 1850–1900* (Bombay and Oxford: Oxford University Press, 1995). The quoted phrase appears in the title of chapter 3 of Kerr's book.

[52] Gilmartin, 'Scientific Empire and Imperial Science', pp. 1127–49.

[53] Klingensmith, *'One Valley and a Thousand'*, chapter 5.

account as key actors in the debates on the nature of electricity generation projects best suited for the Presidency.[54] Ross Bassett's recent book covers a large period over the nineteenth and twentieth centuries, studying a particular group of technical practitioners: Indian graduates of the Massachusetts Institute of Technology.[55]

As this discussion would suggest, historical accounts, like contemporary discourse, use the terms 'engineering' and 'engineer' to denote a wide variety of activities and practitioners at various points in time and in various locations. For the purposes of this study, I have considered as an engineer anyone to whom at least one of the following descriptions applies:

1. Holder of an engineering degree or diploma from an institution in Britain, India, or (in the case of industrial engineers) any other country
2. Member of one or more of the professional engineering societies in London or in India
3. Officer in the Indian PWD or in the engineering or any other technical department (for example, Traffic) of a state-run or company-run railway
4. Royal Engineer employed in one of the services just mentioned
5. Technical expert in a supervisory/managerial position (above, but not including, foreman level) in an industrial enterprise

Independent consulting engineers were few, and rarely feature in this study. The term 'engineer' here is taken to include all branches of engineering—identities crystallized to a large extent around one's bureaucratic (or, in industries, departmental) allegiance.[56]

---

[54] Srinivasa Rao, 'Electrification of Madras Presidency'.

[55] Bassett, *The Technological Indian*.

[56] As is apparent from the definition provided, I have not included engineers of the Post and Telegraphs department of the Government of India (in this book, the term 'Government of India' refers to the colonial government unless otherwise specified), as they were very few in number, choosing to focus instead on the most numerous and prominent groups of government engineers, those in public works and the railways. To illustrate: the technical officers in Telegraphs numbered in the region of forty in 1929 (*India Office List* for 1929), whereas their counterparts in the Public Works Department were several hundred in number, and in the railways more than a thousand (see Chapters 3 and 4).

This wide definition covers individuals with differing speciali-
zations, varying professional interests, assorted nationalities, and
diverse career paths. It would include the irrigation engineer Ganga
Ram (1851–1927), one of the earliest Indians to achieve recognition in
the Public Works Department. A graduate of the engineering college
at Roorkee, Ganga Ram took early retirement from the Punjab PWD
to undertake commercial farming, and later became a philanthropist.[57] It would also include engineers with more modest careers, like
Lala Ram Das (b. 1876), who joined the Punjab Irrigation Service as a
subdivisional officer and rose gradually to the rank of Executive Engineer before retiring in 1931, having also spent part of his service on
secondment to the princely state of Bahawalpur.[58] Of a similar vintage
was J. W. Meares, who straddled the worlds of commerce and government. A Briton who worked in India from 1896 to 1922, Meares held
the posts of Electrical Engineer to the Government of Bengal and later
Electrical Adviser to the Government of India. Unlike many government engineers of his generation, Meares had been trained primarily
by apprenticeship, having been a premium pupil at various works of
the electrical firm Crompton & Co. in Britain; in fact, he originally
came to India as an employee of that firm.[59] This book deals with individuals as varied as R. N. Mookerjee, a civil engineer who co-founded
the contracting firm of Martin & Co. in Calcutta, Cyril Walter Lloyd
Jones, who spent a long and successful career in the employ of the
Nizam's Guaranteed State Railway in the princely state of Hyderabad,
and John L. Keenan, an American steel operator who joined the blast
furnace department of the Tata Iron and Steel Company and rose to
the position of General Manager.[60] Thus, the subjects of this book
include engineers—Indian, European, and American—working in
the governmental, quasi-governmental, and private sectors.

However, this is primarily a book about engineers as a collective, as
members of government services or industrial departments. In places

[57] Mital, *History of the Thomason College of Engineering*, Appendix, p. 262; 'A Great
Indian Enterprise', *Indian Engineering*, 28 March 1925, p. 175.

[58] *India Office List* for 1938.

[59] John Willoughby Meares, 'At the Heels of the Mighty: Being the Autobiography
of "Your Obedient Humble Servant"', typescript (1934), Archives of the Institution of
Engineering and Technology (IET): SC 169/1/1.

[60] See Chapters 2, 4, and 5 of this book.

I study them as individuals, but only when such a focus is relevant to the overall themes of the book.[61] I treat these engineers as belonging to a 'profession'. My aim is not to test whether engineers in India constituted a profession in the strict sociological sense.[62] Instead, I use the concept of profession as a heuristic device that enables me to analyse the interactions between a diverse group of practitioners who nevertheless saw themselves as sharing some important characteristics and having common interests. Indeed, the engineers in this study regularly referred to themselves as professionals, formed professional societies, sought higher status and a sense of community, and portrayed themselves as devoted to the welfare of the public.

In studying engineers in India I have in mind the flexible concept of 'bureaucratic professions' as described by C. W. R. Gispen. Traditional theories of professions, Gispen argues, were developed for an Anglo-American free-market context and modelled closely on medicine and law. Of the several attributes a profession is supposed to possess (esoteric knowledge, autonomy, monopoly of the market, the service ideal),[63] these theories stressed autonomy. Arguing that bureaucrats are not autonomous, they held that bureaucracies and professions were mutually exclusive. Many occupations that are not, by this yardstick, professions (for example, 'the clergy, the officer corps, academics and civil servants') are nevertheless extremely compatible with other attributes supposed to characterize professions, such as 'community, monopoly, closure, service ethic, claims of disinterestedness, and specialized intellectual technique'.[64] It may be

---

[61] For example, the famous engineer M. Visvesvaraya (1861–1962) is not studied in detail, as he was more a technocrat-administrator than a practising engineer in the interwar period. On Visvesvaraya's career as a technocrat, see Dhruv Raina, *Visvesvaraya as Engineer-Sociologist and the Evolution of His Techno-Economic Vision* (Bangalore: National Institute of Advanced Studies, 2001).

[62] Examples of studies that do this for other occupational groups are Paul Brassley, 'The Professionalisation of English Agriculture?', *Rural History*, vol. 16, no. 2 (2005): 235–51; and Jeffery, 'Allopathic Medicine in India'.

[63] On the characteristics of professions, see, for instance, Jan Goldstein, 'Foucault among the Sociologists: The "Disciplines" and the History of the Professions', *History and Theory*, vol. 23, no. 2 (May 1984): 170–92, here p. 175.

[64] C. W. R. Gispen, 'German Engineers and American Social Theory: Historical Perspectives on Professionalization', *Comparative Studies in Society and History*, vol. 30,

argued that the premium placed on the criterion of autonomy is at the heart of the well-known analyses of American engineers by Edwin Layton and David Noble, who view the dual loyalties of engineers to their corporate employers and to the wider profession as fundamentally incompatible.[65] Yet, engineers, a large proportion of whom are employed in governments and private companies, seldom possess complete autonomy.[66] In this respect the experience of continental European engineers in the nineteenth century is particularly relevant to the present study, as a large proportion of them, as in the case of Indian engineers, were civil servants. Whereas Terry Shinn, writing about France, views such engineer-bureaucrats as forming 'corps' as opposed to the 'profession' of industrial engineers,[67] the present study views both types of engineers as belonging to a common profession. This was symbolized by the post–Great War establishment of an all-India professional society in Calcutta, which included government as well as industrial engineers as members (see Chapter 2).[68]

---

no. 3 (July 1988): 550–74; the quote is from p. 556. The term 'bureaucratic professions' appears, for instance, on p. 557n27.

[65] Edwin T. Layton, Jr, *The Revolt of the Engineers: Social Responsibility and the American Engineering Profession* (Baltimore: Johns Hopkins University Press, 1986); David F. Noble, *America by Design: Science, Technology, and the Rise of Corporate Capitalism* (Oxford: Oxford University Press, 1979). See also Peter Meiksins's critique of Layton. Meiksins argues that the 'revolt' described by Layton was not a case of engineers fighting for autonomy from the bureaucratic hierarchies they were a part of; instead, they were agitating for better salary and prospects of promotion *within* the bureaucracies employing them. Peter Meiksins, 'The "Revolt of the Engineers" Reconsidered', *Technology and Culture*, vol. 29, no. 2 (April 1988): 219–46.

[66] See the discussion of Magali Larson's work by Gispen in his 'German Engineers', pp. 555–6.

[67] Terry Shinn, 'From "corps" to "profession": the emergence and definition of industrial engineering in modern France', in *The Organization of Science and Technology in France 1808–1914*, edited by Robert Fox and George Weisz (Cambridge: Cambridge University Press, 1980), pp. 183–208.

[68] A somewhat similar situation is described by Gispen for German engineers in the nineteenth century. He argues that in the Verein deutscher Ingenieure, the major professional engineering institution of the time, bureaucratic and entrepreneurial professionalism balanced each other, keeping the institution in a kind of stasis. Gispen, 'German Engineers', pp. 565–72.

SOURCES AND STRUCTURE

The analysis in this book relies on both quantitative and qualitative research methods. Statistics are compiled to establish the extent of Indianization, the size of the engineering cadres, and the membership of various professional institutions, all of which are important parts of my arguments. The relevant data is compiled from a wide range of sources. These include government reports, for example, annual reports of the Railway Board, reports of specially appointed commissions, and the *Moral and Material Progress* statements submitted annually by the India Office to the British Parliament; records of service of government servants; and topographical membership lists, annual reports, and other publications of British and Indian professional societies. In addition to quantifiable data, these sources, along with others (for example, annual reports and other publications of the Tata Iron and Steel Company; personal papers of individual engineers; obituaries in company magazines, newspapers, and professional journals; official files and correspondence in the India Office Records) are used to build up a picture of engineers' educational qualifications and career trajectories. To explore contemporary ideas about engineers and engineering (engineers' perception of themselves and their social status; policymakers' ideas on race and engineering ability), I study statements by and about engineers in various forums. These include published and unpublished memoirs and biographies; trade magazines such as *Indian Engineering* and the *Indian Railway Gazette*; and debates in the Indian central legislature and in the British Parliament.

Many of the sources mentioned so far were consulted in the Asia, Pacific and Africa Collections (APAC) of the British Library in London. Also important were two as yet underutilized sources on the open shelves of the APAC, the annual *India Office Lists* (brought out by the India Office) and the privately produced almanac, *Thacker's Indian Directory*—both invaluable references for details on official appointments and bureaucratic regulations. Data related to the Institutions of Civil, Mechanical, and Electrical Engineers in London and the Institution of Engineers (India) in Kolkata (formerly Calcutta) were obtained from their respective archives.[69] A rich collection of private

---

[69] The archives of the Institution of Electrical Engineers (IEE) are held at its successor institution, the Institution of Engineering and Technology (IET), London.

papers, administrative files, and correspondence from the interwar period—most of which I believe have not been used before by historians—was examined at the Tata Steel Archives in Jamshedpur. Finally, many sources were accessed via online/digitized archives. These include government reports from the Digital Library of India, online *Proceedings* of the Institution of Civil Engineers and the Institution of Mechanical Engineers, the ProQuest online database of House of Commons Parliamentary Papers, and the online Hansard (transcripts of British parliamentary debates).

Covering the period between 1900 and 1947, but focusing largely on the interwar period, the chapters that follow address developments in the engineering profession across India. (The overall trends refer primarily to the provinces of British India, although the princely states, which had analogous engineering services, and were subject to comparable trends, are also studied where relevant.) As mentioned earlier, these developments are studied in relation to *Indianization* and *industrialization*. The issues represented by these two terms, which became the leitmotifs of policy debates after World War I, are discussed in detail in Chapter 1. Against this backdrop, Chapter 2 analyses, through the lens of professional engineering societies, the interrelationships between different types of engineers and developments in the profession as a whole. It traces the founding (in 1920) and development of the Institution of Engineers (India), the first pan-Indian professional institution for engineers. Through an analysis of membership statistics of this institution and of its older counterparts in London (of which many engineers in India were members), the chapter demonstrates the growing importance of industrial engineers, and the emergence of an Indian identity in a profession that was earlier oriented towards the imperial metropolis.

Chapters 3, 4, and 5 of this book are organized on the basis of the sectors in which engineers were employed: public works, railways, and private industry respectively. I employ this structure advisedly. It is well established that the colonial state's system of engineering education was essentially oriented towards producing public works engineers and subordinates until well into the twentieth century.[70] Yet, this ought not to obscure the importance of railway engineers, who were

---

[70] See Arun Kumar, 'Colonial Requirements'.

present in significant numbers and were organized in bureaucratic services that (especially on the state-run railways) shared marked similarities with the public works engineering service. Indeed, the state railways formed a branch of the PWD until 1905.[71] The analysis of engineers would also be incomplete without taking into account industrial experts, who became increasingly prominent in the inter-war period. As Chapter 2 demonstrates, these engineers shared professional platforms with the government-employed engineers of the public works and railways and often saw themselves as part of the same profession. Yet, their educational backgrounds and career trajectories could be quite different, and Chapter 5 demonstrates this through a detailed study of the pre-eminent Indian steel works (which was fundamental to the growth of industry). These sector-wise studies of engineers, then, illuminate the similarities and differences in their experiences, and illustrate how they were each related to the same broad currents of Indian history.

Chapter 3 studies the reorganization of the engineering cadres in the Public Works Department (PWD) over the years 1900 to 1947. Here, I argue that a PWD engineer was seen first as a gentleman officer and only then as a technical expert; and that colonial officials and British legislators deployed this paradigm when, following the interwar constitutional reforms, they argued for the need to limit the extent of Indianization and the powers of provincial governments in PWD work. In Chapter 4 I qualify the historiographical view that the nationalist movement drove the history of the railways in interwar India. I show that, as in the case of the PWD, the colonial government tried to manage Indianization so as to maintain the existing bureaucratic character of the railways. Extending Daniel Headrick's claim, I demonstrate that such Indianization as did occur was due not only to the recruitment of more Indians, but to the exit of European engineers during the Depression and World War II. Chapter 5, a study of the technical experts of the Tata Iron and Steel Company (TISCO), describes a contrasting process of Indianization, one which was not as dependent on the colonial state or its education system. The company had a multinational team of technical experts, the most senior of them from America, to which was added a steady stream of Indians

---

[71] See Chapter 4.

trained abroad (mostly in the USA; some in Germany and Britain) or in TISCO's in-house training institute, set up in 1921. Thus, the chapter adds a study of technical practitioners to economic historians' analyses of Indian industrialization, which concentrate on factors such as economic protection and the availability of capital. The Conclusion draws together the various strands of analysis from the preceding chapters.

CHAPTER ONE

The Context

Indianization and Industrialization in Indian History

As power is given to the people of a province or Dominion to manage their own local affairs, their attachment becomes the stronger to the Empire which comprehends them all in a common bond of union.

—*Montagu–Chelmsford Report* (1918)[1]

The transition must be accomplished from unresponsive and utterly unresponsible Bureaucracy of to-day, to a gradually more amenable, more understanding and sympathetic Public Service responsible to the chosen Ministers of the popular will.... We must seek to make [the public service] more national and more sympathetic, more understanding of the country's needs and more amenable to its people's wishes.

—K. T. Shah, *Public Services in India* (1935)[2]

[T]he views of Government and of the public have been further modified under the stress of war necessities, which have led to a still more definite adoption of the policy of State participation in industrial development ...

—Report of the Indian Industrial Commission (IIC) (1918)[3]

[1] *East India (Constitutional Reforms). Report on Indian Constitutional Reforms* (Cd. 9109, London: His Majesty's Stationery Office [HMSO], 1918), p. 149.

[2] K. T. Shah, *Public Services in India (Congress Golden Jubilee Brochure—7)* (Allahabad: All India Congress Committee, 1935).

[3] Indian Industrial Commission, 1916–18, *Report* (Calcutta: Superintendent Government Printing, 1918), p. xix.

My particular concern in this book is to examine how developments in the engineering profession were caused by, and in turn influenced, the twin transformations of Indianization and industrialization in interwar India. By reading key primary sources alongside the findings of the historiography, this chapter provides an overview of the course of Indianization and industrialization over the period between 1900 and 1947. This will provide essential context for the subsequent chapters, and serve as a point of departure for the arguments I develop in them about engineers.

THE EARLY HISTORY OF INDIANIZATION

In its literal sense, 'Indianization' in colonial India referred to a bureaucratic process—the replacement of Britons by Indians in legislative bodies, government administration, and private enterprises. But Indianization was always more than this. It was a political idea, a theme central to the negotiations between colonial officials and nationalist politicians in the twentieth century. This is best captured by the words of H. N. Kunzru of the Servants of India Society,[4] who, criticizing the government's attitude to Indianization, wrote in 1917:

> Our vital interests are bound up with the proper solution of this question, which is at once moral, political and economic. It affects our manhood. It involves our national self-respect. It is a test, also, of England's good faith. If she is mindful of her moral responsibilities, if her dominion in India is not to be synonymous with the exploitation of a helpless people, if the Act of 1833 and the Proclamation of 1858 are not mere scraps of paper, it is her bounden duty to raise Indians to positions of trust and responsibility, and to make them feel that they are not treated as helots in their own country.[5]

As this statement suggests, the issue of Indianization had its roots in the nineteenth century.[6] In its Charter of 1833, the East India

[4] The Servants of India Society was a social reformist organization established in 1905 by G. K. Gokhale. See Barbara D. Metcalf and Thomas R. Metcalf, *A Concise History of India* (Cambridge and New York: Cambridge University Press, 2003 [2002]), p. 140.

[5] Hirday Nath Kunzru, *The Public Services in India (Political Pamphlets—II)* (Allahabad: Servants of India Society, 1917), p. 1.

[6] The specific term 'Indianisation'/'Indianization', however, appears to have entered public discourse in the twentieth century. The earliest use of the term I have

Company had declared a policy of encouraging the selection of Indians for government service. When Crown replaced Company in 1858, Queen Victoria proclaimed to the Indian populace that suitably qualified British subjects, irrespective of their race, would be selected to serve the government. In the following decades the scope for employment by the government increased as its administrative bureaucracies grew larger and more specialized. It was also recognized that significant savings could be made by employing Indians, who were paid lower salaries than their British counterparts. Despite all these conditions, a belief in the racial inferiority of Indians—exacerbated by late-nineteenth-century social-Darwinist beliefs—ensured that the government rarely appointed them to responsible positions, one Viceroy even suggesting in private that assurances of doing so were insincere.[7]

Simultaneously, a national consciousness was growing among the urban educated classes. At the annual meetings of the Indian National Congress, formed in 1885, members 'gently reminded' the government of the royal promise of a greater Indian share in government

---

found in primary sources dates back to 1917 in Kunzru's pamphlet (Kunzru, *Public Services in India*, p. 51) and in the testimony of an Indian Civil Service (ICS) officer around the same time, in *Papers Relating to the Application of the Principle of Dyarchy to the Government of India: To Which Are Appended the Report of the Joint Select Committee and the Government of India Act, 1919, with an Introduction by L. Curtis*, edited by L. Curtis (Oxford: Clarendon Press, 1920), p. 179. (This volume is hereinafter cited as Curtis, *Papers Relating to Dyarchy*). The term began to appear regularly in government publications (such as the annual *Statement Exhibiting the Moral and Material Progress and Condition of India*, available in the British Library) in the 1920s. I have been unable to ascertain whether it was a bureaucratic coinage or a term that emerged from the discourse of nationalist politicians. It is more likely to have been the former, as we know that an earlier usage of the term in a very different context (and with a different meaning) was current in British-administered Egypt in the late nineteenth/early twentieth century. There it referred to British administrators' attempts to remodel the Egyptian administration along the lines of the system in British India (Robert L. Tignor, 'The "Indianization" of the Egyptian Administration under British Rule', *The American Historical Review*, vol. 68, no. 3 [April 1963]: 636–61). The label is not merely a retrospective one applied by Tignor. Tignor makes clear on page 661 that 'the term Indianization was familiar to most of these officials [British officials in Egypt]'.

[7] Burton Stein, *A History of India*, 2nd edn, edited by David Arnold (Oxford: Wiley-Blackwell, 2010), pp. 239–43; Metcalf and Metcalf, *Concise History of India*, p. 103.

posts, although their demands were modest and changes were sought largely within the existing constitutional framework.[8] This constitutional system was one in which Indians had no executive powers. Despite the existence of Legislative Councils at the central and provincial levels, there were no elections, and the majority of members nominated were civil servants, most of them Europeans. These 'official' members dominated the executive councils of the Governor-General at the centre and the governors in the provinces, while the Indian members were generally 'non-officials', nominated from among the rest of the public, and had only legislative powers.[9]

Although the Indian National Congress had begun as a group of peaceful petitioners, a more radical view was taken by some sections of the Congress around the turn of the twentieth century. An important stimulus to their radicalization was the unpopular partition of Bengal under Viceroy Curzon in 1905, which was seen as a colonial ploy to split the nationalists in that populous presidency along religious lines. In what has been seen as an effort to wean the Congress away from its radical leaders and keep them with the 'moderate', constitutionalist ones, the post-Curzon administration proposed a set of constitutional reforms (known as the Morley–Minto Reforms), which culminated in the Indian Councils Act of 1909.[10]

Under the new Act, the strength of the central and provincial Legislative Councils was increased to sixty at the centre, and to more than twice the previous size in most provinces. For the first time, Indians were appointed to the executive councils of the Governor-General at the centre and the governors in the provinces. Elections were introduced for some of the seats in the provincial Legislative Councils, although the franchise was limited on the basis of income, and separate electorates were provided for different communities. Although Indians elected to the provincial legislatures had no executive powers—as the Governor's executive council continued to consist of appointees—the proportion of non-official members was now

[8] Stein, *History of India*, pp. 275–6.

[9] See *India Office List* for 1905, pp. 7, 75, for examples of the structure of government at the central and provincial (here Madras Presidency) levels. On official and non-official members, see Metcalf and Metcalf, *Concise History of India*, p. 103.

[10] Stein, *History of India*, chapter 7. The partition of Bengal was revoked in 1911 (Stein, *History of India*, p. 286).

increased to give them a majority in the legislature. (At the centre, there was still an official majority.) Furthermore, members were now allowed to introduce resolutions in the councils as a means of showing their dissatisfaction with government decisions.[11]

Thus, the Morley–Minto Reforms constituted the first notable step in Indianizing the political system. But no appreciable steps had as yet been taken to increase the limited opportunities for Indians in administrative jobs in the government. This was in spite of continuous agitation by Indian politicians for reforms in the composition of the Indian Civil Service (ICS), the predominantly European members of which dominated the most prestigious government appointments. A particular demand was the holding of simultaneous entrance examinations to the ICS in Britain and India, so that Indians would not be obliged to travel to London to take the examination. These demands were mostly ignored, Curzon being particularly intransigent during his tenure as viceroy.[12] Even under existing conditions, he felt that 'higher posts that were meant and ought to have been exclusively and specially reserved for Europeans [were] being filched away by the superior wits of the native in the English examinations'.[13] Yet, the Indianization of the Legislative Councils, which had begun with the Morley–Minto Reforms, began to lead to further changes. In particular, it gave an impetus to Indianization in the administrative services in the interwar years.

CONSTITUTIONAL REFORMS AND THE INDIANIZATION OF PUBLIC SERVICES

The Public Services, 1900–19

In the twentieth century, the terms of the Indianization debate grew to include not only the ICS, but all the public services of the government,

---

[11] *Royal Commission on the Public Services in India: Report of the Commissioners*, vol. I (Cd. 8382, London: HMSO, 1917), p. 16; Stein, *History of India*, pp. 284–5; John F. Riddick, *The History of British India: A Chronology* (Westport, CT: Praeger, 2006), p. 94. The first of these sources is hereinafter cited as *Islington Commission Report*.

[12] Malti Sharma, *Indianization of the Civil Services in British India (1858–1935)* (New Delhi: Manak, 2001), chapter 9.

[13] Curzon to Hamilton, 23 April 1900, quoted in Sharma, *Indianization of the Civil Services*, p. 162. Lord George Hamilton was Secretary of State for India, 1895–1903 (Riddick, *History of British India*, p. 270).

which now depended on a complex set of bureaucracies.[14] Before discussing the Indianization of the services, it will be useful to outline the main features of this bureaucratic set-up in the period 1900–20.

Government administration was carried out through a number of departments. Some of these departments were directly under the central government (foreign affairs, the army, telegraphs, and state-run railways[15]), while others were run by the provincial governments (law and justice, education, income tax, and public works). The officer ranks of the departments were staffed by members of various administrative services, which were 'reservoirs of trained men' created for this purpose.[16] Figure 1.1 shows the different types of departments and services, and the relationships between them.

As the diagram shows, there were three major types of services in early-twentieth-century India: Central, All-India, and Provincial.[17] The Central Services provided the staff required for those departments that operated directly under the central government, for example, telegraphs and state railways. Their members were recruited from two sources: in London, by the Secretary of State's India Office (and in some cases from among fresh Royal Engineer officers); and in India, by the central government. Members of the All-India Services staffed the departments operating primarily in the provinces,

[14] The intensification of bureaucracy had begun after Crown Rule was instituted. Metcalf and Metcalf, *Concise History of India*, p. 103.

[15] This refers to the railways run by the state (usually referred to simply as 'state railways') as opposed to railways managed by private companies. For an explanation of the different kinds of railways, see Chapter 4.

[16] Part IV: 'The Structure of Indian Government (June 1917)', in *Papers Relating to Dyarchy*, edited by L. Curtis. The quoted text appears on page 206.

[17] It is not clear when the distinction between 'All-India Service' and 'Central Service' was first made. Both terms were used in the *Lee Commission Report* (1924), while the term 'All-India Service' was mentioned in the *Montagu–Chelmsford Report* of 1918 (*East India [Constitutional Reforms]. Report on Indian Constitutional Reforms* [Cd. 9109, London: HMSO, 1918]), p. 195 (paragraph 240). For the sake of clarity and continuity, I use the terms 'All-India' and 'Central' here starting from the period between 1900 and 1920: the distinct features defining each were already evident in this period. The full reference for the *Lee Commission Report* is *East India (Civil Services in India): Report of the Royal Commission on the Superior Civil Services in India* (Cmd. 2128, London: HMSO, 1924).

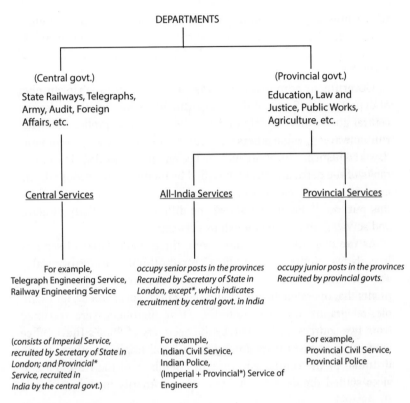

Figure 1.1   Government administration, c. 1900–20: Departments and services

Sources: Compiled by the author from various sources.

Notes: 1. Provincial* is not the same as Provincial Services at the right corner of this diagram.
         'Provincial' in these cases mainly indicates a lower pay-scale than the Imperial Service.
     2. This diagram shows only officers. Subordinate staff were recruited in India.

such as Education and Law (Police). They were recruited entirely by the India Office in London (with one prominent exception in Public Works, which will be discussed later). Members of the Provincial Services were recruited in the respective provinces. They worked in provincial departments alongside members of the corresponding All-India Service, but holding positions of lesser responsibility. While the All-India officers could in theory be transferred to another province or to the central government, provincial officers spent their entire career in one province.[18]

In two cases, those of public works engineers and railway engineers, the term 'Provincial Service' was used in a potentially confusing sense. In state railways as well as the Public Works Department (PWD), the officers recruited in England (who were mostly Europeans) were called the 'Imperial Service', while those recruited by the central government in India were called the 'Provincial Service', composed essentially of Indians, Eurasians, and Europeans permanently resident in India. These Provincial Services were different from the usual ones (such as the Provincial Civil Service) in that the work their members did was in no way different from that of the Imperial engineers; they could also attain the same ranks. Here the word 'provincial' only denoted the fact that they were recruited in India, and that their salary and terms of employment were different from those of the Imperial Service engineers.[19]

There were two primary methods of recruitment to the public services in this period: competitive examinations and 'nomination'.[20] In some prominent services like the ICS and the Indian Police, recruitment of officers in England was carried out through competitive examinations.[21] In others, recruitment was by 'nomination'. In the Engineering Services of the PWD and state railways, for example, the Secretary of State called for applications, and a selection committee was appointed to assess the candidates' qualifications and subsequently invite some for an interview at the India Office, on the basis of which they would be selected. (There were some additional requirements: candidates had to pass a medical examination and a riding test

---

[18] See in particular *Islington Commission Report*, pp. 18–19, and *Lee Commission Report*, p. 3.

[19] *Islington Commission Report*, pp. 18–19.

[20] The reference for the rest of this passage, except where another citation is given, is *Islington Commission Report*, pp. 28–31.

[21] The competitive examination was sometimes combined with nomination, as in the first few years after World War I, when British candidates were scarce in the exam, so that most British recruits were obtained by appointing men who had fought in the war. David Potter, 'Manpower Shortage and the End of Colonialism: The Case of the Indian Civil Service', *Modern Asian Studies*, vol. 7, no. 1 (1973): 47–73, here pp. 51–2.

before they could be confirmed.)[22] Nomination was also the norm in recruiting officers in India, where, with one minor exception, no competitive examinations were held before the end of World War I.[23] In some cases, a system of 'patronage' existed. For instance, Provincial Service engineers for the PWD were selected from the Indian engineering colleges, according to a fixed quota for each college.[24] External recruitment was sometimes supplemented by promotion from a lower service within the department. Thus, in a small number of cases, candidates were promoted from a Provincial Service to the corresponding Imperial Service; and occasionally, subordinate employees were promoted to the Provincial Service in the same department.[25]

The size of the public services varied according to the nature and importance of the associated tasks. In 1913 the sanctioned (that is, prescribed maximum) strengths of some of the services were as shown in Table 1.1.

The organization of the services and the system of recruitment were unfavourable to Indian aspirants in a number of ways. To begin with, as Table 1.1 emphasizes, the prestigious All-India Services were recruited in Britain, which meant that nearly all their members were Europeans. Although Indians were allowed to compete in most cases, few could afford to travel to Britain and acquire the required British qualifications or appear for the competitive examinations.[26] David Potter has calculated that in the years between 1904 and 1913 (both inclusive), 501 out of 528 candidates selected to the ICS through the London examination were Europeans, a proportion of 95 per cent.[27] In

---

[22] See 'Indian Public Works Department: Regulations as to Appointments of Assistant Engineers, 1910' in the *India Office List* for 1910, p. 222; Annexure XVIII: 'Public Works Department and Railway Department (Engineering Establishment)', *Islington Commission Report*, pp. 326–37, here p. 330.

[23] As of the 1910s, the only exception was 'in the Punjab, where out of the total number of candidates annually recruited two are appointed on the results of competitive examination among nominated candidates'. *Islington Commission Report*, p. 29.

[24] Annexure XVIII, *Islington Commission Report*, p. 331; see Chapter 3 of this book for more on PWD recruitment.

[25] *Islington Commission Report*, pp. 19–20.

[26] The difficulty of Indians in going to Britain to compete for Civil Service appointments was already being raised in the British Parliament in the late nineteenth century. See Sharma, *Indianization of the Civil Services*, pp. 143–6.

[27] Potter, 'Manpower Shortage', p. 49.

Table 1.1   Types and sanctioned sizes of public services as of 1913

Name of Service	Type of Service	Strength		
		Recruited in England	Recruited in India	Total
Indian Civil Service	All-India	1,350	61[a]	1,411
Provincial Civil Services	Provincial	–	2,572	2,572
Agricultural Service (Imperial)	All-India	62	–	62
Agricultural Service (Provincial)	Provincial	–	56	56
Forest Service (Imperial)	All-India	213	–	213
Forest Service (Provincial)	Provincial	–	208	208
Police (Imperial)	All-India	661	10[a]	671
Police (Provincial)	Provincial	–	255	255
Public Works Engineering & Railway Engineering	(All-India and Central)	648[b]	280[c]	928
Telegraph Engineering	Central	23[b]	23[c]	46

Key: a: By promotion from Provincial Service; b: Imperial Service; c: Provincial Service ('Provincial' in the sense of India-recruited members of an All-India or Central Service)

Source: Based on table in 'Minute by Mr Abdur Rahim', Royal Commission on the Public Services in India. Report of the Commissioners, Vol. I (Cd. 8382, London: HMSO, 1917) [Islington Commission Report], pp. 394–488, here pp. 395–6.

Note: For confirmation that the figures in the table refer to sanctioned strengths, cf. Annexure XVIII, Islington Commission Report, p. 326, which gives the total sanctioned size of PWD and state railways engineering services together as 928.

the case of the Indian Police, a candidate could only enter the London examination if he was 'a British subject of European descent'. In addition, 'at the time of his birth his father must have been a British subject, either natural born or naturalised in the United Kingdom'.[28]

---

[28] Rule quoted in Annexure XVI: 'Police Department', Islington Commission Report, pp. 297–316, here p. 298.

Further, most Indian officers belonged to the Provincial Services, which implied lower status and a lesser grade of responsibilities than the Imperial officers. It also meant lower pay.[29] This was even true of the case of the exceptional Provincial Service in the PWD and the railways, whose members were engaged in the same tasks as their Imperial Service counterparts and could attain the same ranks. Before World War I, the Provincial Service engineers were paid approximately two-thirds of what the Imperial engineers received, except at Chief Engineer rank, where both were paid equal salaries.[30]

Also contentious was the status of Anglo-Indians and Domiciled Europeans. The term 'Anglo-Indian' at this time usually referred to those of mixed European and Indian ancestry (Eurasians), a community that had grown in India over two or more centuries. 'Domiciled European' referred to those of entirely European parentage but who were permanently resident in India, usually over multiple generations. As historians have shown, these categories (along with the simple 'European') were fluid, always contested, and susceptible to different interpretations, depending on factors such as one's schooling and ownership of property in Britain.[31] In fact, it appears that sometimes *both* categories were implied when reference was made either to 'the domiciled community' or 'the Anglo-Indian community'.[32] Anglo-Indians, Domiciled Europeans, and 'native' or 'Asiatic' Indians together were classed under the rubric of 'statutory natives of India'.[33] The non-Asiatic Indians, so to speak, occupied a large proportion of middle and higher-level jobs in government service. In

[29] See *Islington Commission Report*, p. 19. 'Where there is a large body of work of a less important character to be done ... it would obviously be extravagant to recruit officers to do it on the terms required to obtain men for a higher class of duty.'

[30] See *Historical Retrospect of Conditions of Service in the Indian Public Works Department. (All India Service of Engineers)*, Private Pamphlet (c. 1925) in the Secretary of State's Library Pamphlets, vol. 72, T.724 [APAC: P/T724], pp. 2, 8.

[31] See Elizabeth Buettner, *Empire Families: Britons and Late Imperial India* (Oxford: Oxford University Press, 2004); Laura Bear, *Lines of the Nation: Indian Railway Workers, Bureaucracy, and the Intimate Historical Self* (New York: Columbia University Press, 2007), esp. pp. 299–300 (note 17) and p. 140. See also Frank Anthony, *Britain's Betrayal in India: The Story of the Anglo-Indian Community* (Bombay: Allied Publishers, 1969) on the definition of 'Anglo-Indian' over time.

[32] See 'Minute by Mr W. C. Madge', *Islington Commission Report*, pp. 386–91.

[33] *Islington Commission Report*, p. 23.

1913, they accounted for 1,593 out of 11,064 government posts (14.4 per cent) with a monthly salary of Rs 200 and above, whereas they formed well under 0.1 per cent of the Indian population.[34]

The 'Asiatic' Indians thus felt that they were being excluded from opportunities for government employment even in those branches of the services for which officers were recruited in India. The resentment was largely born of the fact that recruitment in India was by nomination, a system suspected of favouring Anglo-Indians and Domiciled Europeans. Certainly, there was a disparity in the educational requirements stipulated for candidates of different categories. To be eligible for certain subordinate posts, Anglo-Indians and Domiciled Europeans, who were usually educated in schools meant exclusively for them, were only required to complete the 'European schools' curriculum, whereas 'native' Indians had to be university graduates. In the 1910s a government commission proposed a similar system for officer appointments.[35] For their part, Anglo-Indians and Domiciled Europeans often expressed dissatisfaction that their terms and share of employment were not on a par with that of the 'pure' Europeans.[36]

Expanded Legislatures and the Islington Commission (1912–15)

Indian politicians were vocal in their protests against the disadvantaged position of Indians in the public services. At its annual meetings, presidents of the Indian National Congress reiterated the demand for reform in ICS recruitment, while in the newly enlarged Imperial Legislative Council, the Congress Moderate Gopal Krishna Gokhale was particularly active in raising the issue of Indian representation

[34] *Islington Commission Report* (table on p. 24); 'Minute by Mr M. B. Chaubal', *Islington Commission Report*, pp. 373–86, here p. 379. The Islington Commission's figures cited here were for 'Anglo-Indians', which I have taken to include Domiciled Europeans as no separate figures were reported for that category.

[35] On the requirements for subordinate posts, see Buettner, *Empire Families*, p. 91. For officer positions, Anglo-Indians and Domiciled Europeans would have to pass 'an examination of a corresponding standard [to the university degree] in the European schools course'. Quoted in 'Minute by Mr M. B. Chaubal', p. 381.

[36] 'Minute by Mr W. C. Madge'; *Historical Retrospect of Conditions of Service in the Indian Public Works Department*. The latter was a memorial to the Secretary of State from Domiciled European engineers in the PWD, asking to be given the same terms of employment as Europeans recruited in Britain.

in the public services. In 1911 another Indian member of the central legislature, N. Subba Rao, moved a resolution asking that a commission be appointed to consider the question. Other Indian legislators supported the resolution. In addition, according to Malti Sharma, the then viceroy, Lord Hardinge (1910–16), was favourably disposed towards Subba Rao's request, which was granted.[37]

As a result, the Royal Commission on the Public Services in India (Islington Commission) was set up in 1912. The Commission had twelve members, three of whom were Indians (including Gokhale, who died in 1915 before the Commission's report was published). Of the remaining nine, at least one was Anglo-Indian, and the rest British.[38] Its task was to study the public services and make recommendations regarding the procedure for recruitment and training of officers; their pay and emoluments; and, most importantly, '[s]uch limitations as still exist in the employment of non-Europeans and the working of the existing system of division of services into Imperial and Provincial'.[39] The Commission carried out its study over the next three years, but because of wartime delays, the report was only published in 1917. Their main recommendations on the subject of Indianization were as follows:

First, they addressed the issue of the misleadingly named Provincial Engineering Services in the state railways and the Public Works Department. In both these departments the distinction between Imperial and Provincial Services was to be abolished by merging the two into a single service.[40]

Second, the Commission recommended differing combinations of recruitment in Britain and India for different types of public service. Indian Finance and Military Finance could henceforth be filled entirely by recruitment in India—provided that there were 'no military

---

[37] Sharma, *Indianization of the Civil Services*, chapters IX and X; 'Subba Rao, Nyapathi' in *The Indian Biographical Dictionary 1915*, edited by C. Hayavadana Rao (Madras: Pillar & Co., 1915), pp. 419–20. For more detail on the arguments made in this debate, see 'Minute by Mr Abdur Rahim', in *Islington Commission Report*, pp. 394–488, here pp. 394–5.

[38] *Islington Commission Report*, pp. 2, 7; W. C. Madge identifies himself as an Anglo-Indian in 'Minute by Mr W. C. Madge'.

[39] *Islington Commission Report*, p. 2.

[40] *Islington Commission Report*, pp. 18–19.

considerations to the contrary' in the latter department. The bulk of recruitment should continue to be made in England for the ICS and Indian Police, 'in which it should be recognised that a preponderating proportion of the officers should be recruited in Europe',[41] although provision was made for the recruitment of a small number of officers in India. (No reason was provided as to why this should be the case—although it probably reflects a policy of retaining responsibility for security in European hands. A similar concern seems to have been behind the old rule, under which recruitment in London for the Indian Police was restricted to European candidates.)[42] In other services, such as those pertaining to Education, Medical, and Public Works, there were 'grounds for policy'—again unspecified—'for continuing to have, in the personnel, an admixture of both western and eastern elements'. Here, a mixture of recruitment in Britain and India was to be adopted: proportions were specified for each department, for example, 50–50 in Public Works.[43] Where the main reason for recruitment in Europe was the lack of technical training facilities in India (for example, Agricultural Service, Geological Survey), such facilities should be built up with the goal of recruiting entirely in India.[44]

Third, the Commission was against the institution of competitive examinations in India, arguing that the education system in the country was not yet sufficiently developed, nor uniformly distributed across regions, for examinations to yield the required type and mix of officers. The argument was that competitive examinations were only suitable in England because of the general educational standard prevailing there. The success of the exams for the Indian Civil and Medical

---

[41] *Islington Commission Report*, p. 61 (paragraph vi).

[42] For the ICS, nine officers annually were to be recruited in India on the basis of a competitive exam but with 'safeguards' (Annexure X: 'Indian and Provincial Civil Services', *Islington Commission Report*, pp. 161–239, here pp. 171–2, paragraphs 18–20). In the case of the Indian Police, the Commission recommended that the bar on Indians competing in London be removed. Further, an existing provision, under which the governor-general could make some appointments in India from among Domiciled Europeans and Anglo-Indians, should be extended to include 'native' Indians too (Annexure XVI: 'Police Department', *Islington Commission Report*, pp. 297–316, here pp. 298–9).

[43] Quoted text from *Islington Commission Report*, p. 61 (paragraph vi); see also p. 27 (paragraph 36).

[44] *Islington Commission Report*, p. 61 (paragraph vi), p. 22 (paragraph 32).

services (held in London) and of the British civil service exams was cited in support of this view.[45] While not endorsing new competitive examinations in India, the Islington Commission suggested some measures to improve the nomination process there—and increase the proportion of Asiatic Indians—such as creating selection committees with Indian members, advertising openings systematically, and starting new technical institutions.[46] (It should be noted here that in 1922 an ICS exam was instituted in India, but—primarily because it was held *after* the London exam each year—it remained of limited importance as a source of Indian officers.)[47]

Significantly, although recommendations were made about the place of recruitment, the Islington Commission declined to use the method of race-based quotas, 'mainly because we recognise the tendency of a minimum to become a maximum, and wish to establish nothing which will prevent qualified Indians, where available, from being appointed in any number on their merits'.[48] This approach was also taken in some cases—notably for the state railways—by the next important Royal Commission, as discussed later. However, it was not endorsed by Indian opinion, which found the recommendations on Indianization inadequate, the Allahabad-based *Leader* making the point that the Commission had concentrated on the pay and emoluments of officers to the detriment of the more important issue of Indianization.[49] Indian politicians and activists, while acknowledging that the recommendations would go some way to ameliorate the position of Indians, were sharply critical of some of the principles underlying the report. Particular exception was taken to the Commission's stand on maintaining in the ICS and Indian Police a 'preponderating

[45] *Islington Commission Report*, pp. 29–33. In fact, competitive examinations had only accounted for one-third of appointments to the British civil service between 1871 and 1910, the rest being filled through patronage or by candidates with specialized professional training (Rodney Lowe, *The Official History of the British Civil Service: Reforming the Civil Service*, vol. 1: *The Fulton Years, 1966–81* [Abingdon: Routledge, 2011], p. 21).

[46] *Islington Commission Report*, p. 62 (paragraphs ix and xi), p. 27 (paragraph 36).

[47] Potter, 'Manpower Shortage', esp. pp. 51 and 55, including Tables 1 and 2.

[48] *Islington Commission Report*, p. 26.

[49] The reaction of Indian newspapers is described by Shukla, *Indianisation of All-India Services*, pp. 200–3. He refers to the view of the *Leader* on p. 201. See also 'Minute by Mr M. B. Chaubal'; and Kunzru, *Public Services in India*, esp. p. 175.

proportion of officers ... recruited in Europe'. Indian members of the Imperial Legislative Council criticized this stipulation as it applied to the ICS,[50] while H. N. Kunzru of the Servants of India Society saw in it a thinly veiled form of racial discrimination: 'The doctrine it lays down implies that the principle of British sovereignty carries with it the corollary that power must for ever be wielded by men of British birth, and teaches Indians that the maintenance of British rule is incompatible with their highest development.'[51] Kunzru further pointed out that some of the Commission's points regarding the need for Europe-recruited officers were pure assertions rather than arguments, and that the evidence placed before the Commission did not provide any grounds for its stance in these cases.[52]

Another criticism was of the Islington Commission's position on race-based quotas. The Commission had been specifically charged with looking into the prospects of 'non-Europeans'—which it defined as 'Indians and Burmans of unmixed Asiatic descent'[53]—but its recommendations, as we saw, only dealt with place of recruitment (in England or in India). In this they were seen as evading another issue—the low proportion of non-Europeans even among officers recruited in India. Indeed, one Indian member of the Commission, M. B. Chaubal, argued in a dissenting minute that '[t]he very meagre percentage of the Asiatic Indians in the higher service ought not to be hidden from view by lumping the Anglo-Indians and the Asiatic Indians together, under the plausible excuse of the definition of "statutory natives of India"'.[54]

Some of the recommendations of the Islington Commission were put into practice, most notably in 1920, when an amalgamated Indian Service of Engineers in the Public Works Department and an analogous service for the state railways were created (see Chapters 3 and 4). But the rest of the report, especially the parts dealing with percentages to be recruited in India, was soon made redundant when further

---

[50] Sharma, *Indianization of the Civil Services*, pp. 198–9.

[51] Kunzru, *Public Services in India*, p. 173.

[52] See Kunzru's points on Public Works, Railways and Telegraphs in *Public Services in India*, pp. 149–51.

[53] *Islington Commission Report*, p. 23.

[54] 'Minute by Mr M. B. Chaubal', p. 380. See also p. 381 (paragraph 24).

changes were made in response to political developments occurring
during and after World War I.

The 1919 Reforms and the Lee Commission (1923–4)

World War I brought the issue of Indian participation in government
councils and services to the forefront of policy debates. Historians
have identified several factors in this process. First, Indian politicians
had cooperated with the war effort, actively campaigning to raise
funds and recruit men for the army. The large number of Indians
who thus fought alongside British and dominion soldiers in Europe
and Mesopotamia created a new feeling of confidence among Indians. In these circumstances Indians were in a position to press for
reforms in return. Second, the Allied Powers' rhetoric emphasized
that they were on the side of freedom against a belligerent Germany,
and invoked the right of peoples to self-determination. The British
government could not logically deny that this applied to the Indian
people too. Third, there was a temporary unity among Indian politicians as they pressed the case for Indian self-government. In particular, the Congress cooperated with the Muslim League (which had
been created a decade earlier, at the time of the Bengal partition) in
1916 to place before the government a joint resolution, asking for a
definite statement from the British monarch that 'self-government'
in India would be introduced as swiftly as possible. These incremental changes were accompanied by other more radical developments.
Some revolutionaries sought to use violent means to overthrow the
colonial government, and repressive laws were enacted. Meanwhile,
Annie Besant (head of the Theosophical Society in Madras) and Bal
Gangadhar Tilak began Irish-inspired Home Rule Leagues, demanding immediate self-government as opposed to incremental reforms.[55]

Responding to the demands of the moderate Indian politicians
who had supported the war effort while attempting to weaken the
hand of radical nationalists, the government proposed a new policy of

---

[55] Stein, *History of India*, chapter 7; Shukla, *Indianisation of the All-India Services*,
pp. 203–6; Metcalf and Metcalf, *Concise History of India*, chapters 5 and 6; Riddick,
The History of British India: A Chronology, p. 96; Curtis, *Papers Relating to Dyarchy*, pp.
359–66 (the Congress–League demand on 'self-government' is quoted on pp. 359–60).

constitutional reform.[56] On 20 August 1917, in a famous statement in the House of Commons, Edwin Montagu, the Secretary of State for India, announced that the British government's 'policy ... is that of the increasing association of Indians in every branch of the administration, and the gradual development of self-governing institutions, with a view to the progressive realisation of responsible government in India as an integral part of the British Empire'.[57] Montagu then proceeded to India at the behest of the Viceroy, Lord Chelmsford, to discuss how this policy might be implemented. After much public debate, the process culminated in the Government of India Act of 1919.[58]

The Montagu–Chelmsford Reforms, as the provisions of the 1919 Act were known, were introduced as a step towards 'responsible government'. A bicameral legislature was created at the centre, with a Legislative Assembly and a Council of State, both of which had a majority of seats reserved for elected members (as against nominated members). In the provinces, too, the councils were given an elected majority, and elected members given a small measure of executive power in a system known as 'dyarchy'. Under dyarchy, fields of administration applicable to the provinces were divided into two categories:

1. 'Transferred' fields, which were placed under the charge of elected provincial ministers—these included Agriculture, Education, PWD (Roads and Buildings); and
2. 'reserved' fields, which were retained under the provincial Governor and his executive council, representing the Crown—for example, PWD (Irrigation), Law and Order, Revenue.

In summary, 'responsible government' was introduced on a small scale, with the paternalistic promise that the situation would be reviewed in ten years' time and the role of ministers expanded if they

---

[56] Burton Stein writes that the new policies were 'intended to broaden the strata of Indians willing to collaborate with imperial authority'. Stein, *History of India*, p. 293.

[57] 'India (Government Policy)', HC Debate 20 August 1917, Hansard (fifth series, vol. 97), cc. 1695–7, here c. 1695. Via ProQuest, House of Commons Parliamentary Papers Online.

[58] 'India (Government Policy)', HC Debate 20 August 1917, Hansard (fifth series, vol. 97), c. 1696; Curtis, *Papers Relating to Dyarchy*, esp. 'Introduction'.

were deemed to have administered their areas capably in the inter-
vening period. Important caveats were built into the reforms too: the
viceroy had a power of veto over the laws drafted by groups with a
majority in the legislatures.[59]

The import of the new policy for the services was twofold. First,
Montagu's reference to 'the increasing association of Indians in every
branch of the administration' was naturally understood by Indian
leaders as a commitment to further Indianize the services. Second,
the introduction of dyarchy made the position of the All-India Services ambiguous. Indian politicians felt that the presence of officers
recruited by, and responsible to, the Secretary of State in services that
were operating in transferred fields necessarily curtailed the autonomy of elected ministers in the provinces.[60]

A third factor also came into play: the Indian public services were
becoming less attractive both to British officers already in India and
to potential applicants in Britain. They were apprehensive about their
status under the changing political system. They were also wary of the
political disquiet that was growing in India,[61] mainly as a reaction to
draconian anti-sedition measures introduced towards the end of the
war, such as the infamous Rowlatt Act of March 1919, which perpetuated the government's wartime extraordinary powers by allowing for
detention without trial for the next three years. After the massacre at
Amritsar the following month, when General Dyer had ordered soldiers to open fire on hundreds of demonstrators in an enclosed park
called Jallianwala Bagh, 'the reforms had become a poisoned chal-
ice'.[62] Meanwhile, Gandhi had begun the non-cooperation movement
in 1920.[63] In addition to these developments, the cost of living in

[59] Stein, *History of India*, chapter 7, esp. p. 293; *Lee Commission Report*, p. 7;
Curtis, *Papers Relating to Dyarchy*; *East India (Constitutional Reforms). Report on Indian
Constitutional Reforms* (Cd. 9109, London: HMSO, 1918), p. 213 (paragraph 263). The
last of these sources is hereinafter cited as *Montagu–Chelmsford Report*. See also *India
Office List* for 1929 (pp. 7–12, 43) for an example of the structure of government that
came into place at the central and provincial levels after the Montagu–Chelmsford
Reforms were introduced.

[60] *Lee Commission Report*, pp. 5–6.

[61] *Lee Commission Report*, p. 6.

[62] Riddick, *History of British India*, pp. 102–3; quoted text from Metcalf and Metcalf,
Concise History of India, p. 165.

[63] Stein, *History of India*, chapter 7.

India had gone up in the aftermath of the war. A particular grievance was that as the rupee's value relative to the pound had fallen, British officers' expenses incurred in importing British goods and educating their children in Britain had, in effect, risen.[64]

This three-pronged problem with regard to the services—the implementation of Indianization, the status of the All-India Services under dyarchy, and the Secretary of State's need to attract British applicants—led in 1923 to the institution of a fresh commission: the Royal Commission on the Superior Civil Services in India (Lee Commission), with five British and four Indian members.[65]

The Lee Commission recommended the provincialization of the All-India Services in fields that had been transferred to elected ministers under the Montagu–Chelmsford Reforms (for example, roads and buildings). This meant that all future recruitment for posts in transferred departments would be made by the concerned provincial ministers—thus addressing the complaint that officers responsible to London could not logically serve ministers responsible to Indian voters.[66] However, existing All-India officers would not be removed from their provincial posts. They were given the option to stay on under their old terms of employment (that is, 'retain their All-India status'), to relinquish their old terms and sign fresh contracts with the provincial government they were serving, or to retire early with 'proportionate pension'.[67] This change also implied Indianization in the transferred fields, as officers recruited in the provinces would most likely be Indians.[68]

As for Indianization of the services in fields reserved to the control of provincial governors (for example, revenue, law and order), the Commission introduced quotas for recruitment for Europeans and Indians (and in some cases by place of recruitment), superseding the recommendations of the Islington Commission. For instance, in

[64] *Lee Commission Report*, pp. 5, 24–5. According to David Potter, writing on the ICS in particular, potential recruits were discouraged more by unattractive pay and service conditions than by the political situation. Potter, 'Manpower Shortage', p. 53.

[65] *Lee Commission Report*, pp. 5–6; see p. ii for composition of Commission.

[66] *Lee Commission Report*, p. 62 (paragraphs i, ii, and iii); Shukla, *Indianisation of All-India Services*, p. 255.

[67] *Lee Commission Report*, p. 70 (paragraph xliii).

[68] Shukla, *Indianisation of All-India Services*, p. 255.

the ICS, 40 per cent of new recruits were to be Europeans, 40 per cent Indians, and the remaining vacancies to be filled by promoting Indians from the Provincial Civil Service—a proportion calculated to bring about within fifteen years an overall 50–50 composition of Europeans and Indians in the ICS.[69] For the Indian Police the time allowed for the achievement of a 50–50 composition was twenty-five years.[70] In the case of the Irrigation branch of the PWD (recruitment for which remained in the hands of the Secretary of State under dyarchy) no such target was set, but the Commission recommended that fresh recruitment be carried out in the ratio of 40:40:20 (European/Indian/Indians promoted from Provincial Services).[71] For the Superior Services of the state-run railways (which formed Central Services), the Lee Commission recommended a mix of recruitment in India and Britain. The prescribed ratio here was 3:1, that is, 75 per cent of vacancies were to be filled by officers recruited in India. However, the government was allowed some latitude: the 3:1 ratio was to be implemented not forthwith but 'so soon as practicable'.[72] (The official response to the recommendations on railways, and the reasons behind it, are analysed in Chapter 4.)

Finally, to improve the financial situation of British officers, the Commission increased their overseas pay. Officers of 'non-Asiatic domicile' were also permitted, once they had completed four years of service, to remit their salaries (earned in rupees) to England at a fixed exchange rate of 2s. to the rupee, which was more favourable than the actual rate.[73]

As J. D. Shukla has shown, Indian opinion, as represented in the country's newspapers, was dissatisfied with the recommendations. They found the proposed rate of Indianization too conservative. They considered the measures regarding the emoluments of European officers a drain on national resources:[74] some Indians called the additional cost of the new measures 'Lee Loot'.[75] In the Legislative

---

[69] *Lee Commission Report*, p. 65 (paragraph xiii). The report does not give the percentage of Indians in the ICS as of 1924.

[70] *Lee Commission Report*, p. 65 (paragraph xiv).

[71] *Lee Commission Report*, p. 65 (paragraph xvi).

[72] *Lee Commission Report*, p. 66 (paragraph xvii-d).

[73] *Lee Commission Report*, p. 67 (paragraphs xix and xx) and pp. 24–5.

[74] Shukla, *Indianisation of All-India Services*, p. 259.

[75] Potter, 'Manpower Shortage', p. 54.

Assembly in Delhi, the government's resolution to adopt the Lee recommendations was defeated, with Motilal Nehru arguing instead for a complete cessation of recruitment in England. Nevertheless, the presentation of the report to the assembly was a mere courtesy—after all, it was concerned with officers recruited by the Secretary of State. The assembly was overruled by the British Parliament, and the Government of India (Civil Services) Bill became law in January 1926.[76]

Further Provincialization: The Government of India Act, 1935

The 1919 Government of India Act had provided for a review of its working at the end of ten years. The Simon Commission was appointed in 1927 to carry out this review, but its all-British composition angered Indian politicians. Eventually, Indian representatives were invited to a series of Round Table Conferences in London in the early 1930s to discuss further constitutional reform, culminating in the Government of India Act of 1935.[77]

The 1935 Act introduced provincial autonomy in all fields, except those of strategic importance (for example, defence, external affairs), of which the viceroy retained direct control. At the centre, a federal council of representatives from the provinces and princely states was to be established; but the federation was to come into being only when 50 per cent of the princes had accepted it, a condition which was never fulfilled.[78] Historians have argued that the 1935 Act, drafted and passed under a national government in Britain in which Conservatives had a powerful say, was designed to limit the influence of

[76] Shukla, *Indianisation of All-India Services*, pp. 260–1; *Statement Exhibiting the Moral and Material Progress of India* for 1925–6, p. 4. See also the discussion of the Lee Commission's recommendations in the House of Lords ('Government of India [Civil Services] Bill', House of Lords Debate 60, cc. 915–28 [Hansard], available at http://hansard.millbanksystems.com/).

[77] Metcalf and Metcalf, *Concise History of India*, chapter 6; R. J. Moore, 'The Problem of Freedom with Unity: London's India Policy, 1917–47', in *Congress and the Raj: Facets of the Indian Struggle 1917–47*, edited by D. A. Low, 2nd edn (New Delhi: Oxford University Press, 2004), pp. 375–403; Andrew Muldoon, *Empire, Politics and the Creation of the 1935 India Act: Last Act of the Raj* (Abingdon, Oxon: Ashgate, 2009), chapter 2; Riddick, *History of British India*, pp. 106–10; Stein, *History of India*, chapter 8.

[78] Stein, *History of India*, p. 326; Metcalf and Metcalf, *Concise History of India*, pp. 192–3; Moore, 'The Problem of Freedom with Unity'.

the Congress, and to retain ultimate control of India in British hands. Through the federal government at the centre—where a large proportion of seats would be reserved for Muslims and the princely states—the new constitution would confine the influence of Congress to the provinces.[79] As Lord Linlithgow (Viceroy, 1936–43; Chairman of the 1935 Joint Parliamentary Committee that considered the Government of India Bill) wrote privately some years later, the 1935 constitution was designed to

> [maintain] British influence in India. It is no part of our policy, I take it, to expedite in India constitutional changes for their own sake, or gratuitously to hurry the handing over of the controls to Indian hands at any pace faster than that which we regard as best calculated, on a long view, to hold India to the Empire.[80]

While the federation did not materialize, provincial autonomy came into effect. Elections were held in 1937 with an enlarged electorate of around 30 million voters; the Congress won around 50 per cent of all seats and assumed power in several provinces.[81] The All-India Services which had not already been provincialized were now placed directly under the elected ministers, although British parliamentarians resisted the move (primarily as it meant the handing over of more responsibilities to Indian officers).[82] Exceptions were made, however, for the ICS, the civil branch of the Indian Medical Service, and the Indian Police, which remained in the hands of the Secretary of State. However, all existing All-India officers continued on their old terms, so that the old cadres were not abolished but allowed to dwindle as their members retired.[83] The 1935 Act also made several provisions to

[79] Moore, 'The Problem of Freedom with Unity'; Metcalf and Metcalf, *Concise History of India*, chapter 6.

[80] Quoted in Moore, 'The Problem of Freedom with Unity', p. 379.

[81] Metcalf and Metcalf, *Concise History of India*, p. 193.

[82] I analyse the arguments advanced for and against the measure in the case of the Irrigation Service, in Chapter 3.

[83] *Government of India Act 1935* (26 Geo. 5. Ch. 2), pp. 148–9 (paragraph 244); Shukla, *Indianisation of All-India Services*, p. 335. Shukla mentions the Indian Agricultural Service, Indian Service of Engineers and Indian Educational Service among the services fully provincialized at this stage.

protect them from the authority of elected Indian ministers.[84] These measures aimed at strengthening the position of officers recruited in England are best understood in the light of David Potter's argument that the British and colonial Indian governments, unwilling to entrust Indian officers with what they saw as crucial responsibilities, relied heavily on the continuing availability of British officers in order to retain control of India.[85] Later in this book, I extend Potter's line of argument to show the particular nature of such concerns (and the reasons behind them) as expressed by officials with regard to public works and railway engineers.

As a result of the negotiated process of Indianization, the major services arrived approximately at a 50–50 composition of Indians and Europeans by around 1940. The ICS, for instance, had 597 Indians and 588 Europeans as of 1 January 1940,[86] while the railways' Superior Services were 50 per cent Indian by 1939.[87]

INDUSTRIALIZATION: INTERWAR POLICY AND THE GROWTH OF LARGE-SCALE INDUSTRY

Just as World War I and the ensuing political changes had important consequences for government services, they also influenced the performance of existing industries and the growth of new ones. The war provided a boost to industrial production as imports fell and the demand for Indian goods increased.[88] More importantly, the war marks the point when Britain's economic policies in India underwent an important shift. In the nineteenth century, the colonial government

[84] See the critique by K. T. Shah, *Public Services in India (Congress Golden Jubilee Brochure—7)* (Allahabad: All India Congress Committee, 1935).

[85] Potter, 'Manpower Shortage'.

[86] T. H. Beaglehole, 'From Rulers to Servants: The I.C.S. and the British Demission of Power in India', *Modern Asian Studies*, vol. 11, no. 2 (1977): 241.

[87] Ian J. Kerr, *Engines of Change: The Railroads That Made India* (Westport, Connecticut, and London: Praeger, 2007), p. 119. Kerr appears to be referring here to state-run and company-run railways together. See also Chapter 5 of this book for a detailed analysis of Indianization in the railways, especially in the interwar period.

[88] Tirthankar Roy, *The Economic History of India 1857–1947*, 2nd edn (New Delhi: Oxford University Press, 2006), pp. 228–9; R. K. Ray, *Industrialization in India: Growth and Conflict in the Private Corporate Sector, 1914–47* (New Delhi: Oxford University Press, 1982 [1979]), p. 4.

had declined to play an active role in the promotion of industry in India or to afford it any protection, thereby safeguarding the interests of British industries that exported their products to India. But during the Great War, when resources, technical personnel, and technical equipment were diverted from India, the dependent position of the Indian economy was exposed. The government saw that the lack of industrial capability had adverse implications for India's security, for its ability to contribute to the war effort in Europe (except in the form of soldiers for the army), and for the stability of the imperial economic system in the face of foreign competition. In May 1916, the Government of India's Department of Commerce and Industry issued a resolution declaring that 'the time has come when the question of the expansion and development of Indian manufactures and industries should be taken up in a more comprehensive manner than has hitherto been attempted'. The resolution announced the appointment of what became known as the Indian Industrial Commission (IIC) to investigate the situation and suggest ways to promote industry in India.[89]

The IIC, presided over by Sir Thomas Holland, a former director of the Geological Survey of India,[90] made a number of important recommendations in the report it published in 1918. Among other measures, the Commission suggested that provincial governments should provide technical assistance to select industries and support private industrialists in providing artisanal/industrial education. Industrial banks should be created to make capital available to industrialists,

[89] Daniel R. Headrick, *The Tentacles of Progress: Technology Transfer in the Age of Imperialism, 1850–1940* (New York and Oxford: Oxford University Press, 1988), p. 338; Aparna Basu, 'Technical Education in India, 1854–1921', in *Essays in the History of Indian Education* (New Delhi: Concept, 1982), pp. 39–59, here pp. 54–5; 'The Indian Industrial Commission: Its Report Summarised', in *East India (Industrial Commission): Report of the Indian Industrial Commission, 1916–18* (Cmd. 51, London: HMSO, 1919), pp. 1–4. The quote is from the Government of India Resolution dated 19 May 1916, printed as Appendix A-1 in the last named source (Cmd. 51, 1919), pp. 301–2, here p. 301.

[90] L. L. Fermor, 'Thomas Henry Holland, 1868–1947', *Obituary Notices of Fellows of the Royal Society*, vol. 6, no. 17 (November 1948): 83–114, here p. 86; Roy M. MacLeod, 'Holland, Sir Thomas Henry (1868–1947)', *Oxford Dictionary of National Biography*, Oxford University Press, 2004 (available at http://www.oxforddnb.com/view/article/33945, accessed 11 June 2012).

and in some cases the central government might also assist industries financially. In order to discharge all these functions, departments of industries should be created in every province. Further, the IIC recommended that an Imperial Department of Industries be created at the centre, and an Imperial Industrial Service be constituted to staff the departments of industries throughout India. The Commission also stated that materials and machinery required for government and railways should, whenever possible, be purchased in India. Provincial Stores officers could also consider buying supplies directly from other provinces.[91] This last constituted a significant change from the government's existing policy, under which stores were purchased exclusively in London through a Stores Purchase Department. Although the policy of purchasing in India was inconsistently applied over the following decades, the government became one of the most important customers of Indian industries.[92] After 1919, the responsibility for industries devolved upon provincial ministers, but many of the IIC's recommendations applicable to the provinces remained relevant. Bengal, for instance, created a provincial Department of Industries in 1920.[93] Although the IIC's vision could not apply in its entirety to post-1919 India, the main principles underlying its report made the Commission an important step in the evolution of the government's industrial policy. As Shiv Visvanathan has argued, the IIC's report formed 'a complete grid on which all the later debates on science and technology may be plotted'.[94]

A related and equally important factor determining the fate of India's industries was the changing fiscal relationship between India and Britain. The Montagu–Chelmsford Report (1918) approved economic protection in principle, and the Indian Fiscal Commission (1921–2) sanctioned the raising of protective tariffs against imports, albeit

[91] Indian Industrial Commission, 1916–18, *Report* (Calcutta: Superintendent Government Printing, 1918), chapter XXIV, 'Summary of Recommendations'. This source is hereinafter cited as IIC *Report*.

[92] Dietmar Rothermund, *An Economic History of India: From Pre-Colonial Times to 1991*, 2nd edn (London: Routledge, 1993), pp. 116–17.

[93] A. Z. M. Iftikhar-ul-Awwal, *The Industrial Development of Bengal: 1900–1939* (Delhi: Vikas Publishing House, 1982), p. 59.

[94] Shiv Visvanathan, *Organizing for Science: The Making of an Industrial Research Laboratory* (New Delhi: Oxford University Press, 1985), p. 42.

with some caveats. To reduce the burden on the Indian consumer, 'discrimination' was to 'be exercised in the selection of industries for protection', which meant that only those industries would be protected which enjoyed 'natural advantages' (such as access to raw materials and labour), which would not be able to develop in the absence of protection, and which showed the potential to compete in time with foreign manufactures in an open market. To decide which industries should be granted protection, an Indian Tariff Board was to be set up.[95]

The policy of protection was clearly related to the growth of several industries in the 1920s and 1930s. In this period, fifty-one enquiries were made by the Tariff Board, and eleven industries were granted protection, including iron and steel, cotton textiles, paper, sugar, and heavy chemicals. These industries grew, mainly by producing substitutes for imported products.[96] With the help of protection, many industries also rode out the Great Depression successfully. While enterprises mainly exporting their products (for example, jute manufacturers) experienced difficulties,[97] conditions during the Depression were actually beneficial to the steel industry (based on import substitution), as any loss in demand had to be borne by imports on which tariffs were payable.[98]

There was also a change in the ownership pattern of industries in the interwar period. The older European managing agencies receded in importance—their business was founded upon exports, and therefore did not benefit from protection. On the other hand, foreign manufacturers saw the opportunity to invest in industry within India, with a view to capturing the (protected) domestic market. Thus, British, European, and American firms set up subsidiaries in India (referred to as 'multinationals') such as Imperial Chemical Industries (ICI), Lever Brothers, and the Swedish Match Company (Wimco). But the most important development was the rise of Indian business houses, which increasingly invested in large-scale industries. These had their

[95] 'Summary of Recommendations', *Report of the Indian Fiscal Commission, 1921–22* (Simla: Superintendent Government Central Press, 1922), pp. xv–xvii (includes quoted text); *Report of the Indian Fiscal Commission, 1921–22*, chapters 1 and 7.

[96] T. Roy, *The Economic History of India*, pp. 229–30; Rothermund, *An Economic History of India*, p. 110.

[97] T. Roy, *The Economic History of India*, p. 231.

[98] R. K. Ray, *Industrialization in India*, p. 77.

origins in Indian communities which, by tradition, had been associated with trade. Most of them had made vast profits through speculation during World War I, enabling them to enter the industrial sector in the interwar period. In some cases these business families started out by manufacturing products covered by the tariffs, such as sugar and paper, then reinvested their profits in more specialized industries such as shipping and sewing machines. Indian businessmen also developed contacts with the Indian National Congress, and grew increasingly aware of their collective interests. This led to the formation (in 1927) of the Federation of Indian Chambers of Commerce and Industry (FICCI) to counter the Associated Chambers of Commerce, which mainly represented the interests of British firms in India. The importance of Indian-owned companies increased. They were responsible for the lion's share of fresh private investment in industry in the interwar years, particularly in the 1930s. By 1944 these companies also accounted for over 80 per cent of workers employed in large-scale industry in the country.[99] R. K. Ray has advanced two explanations for this Indianization of capital investment in industry: first, some Indian businesses were bold in their investments as they were guided by a spirit of economic nationalism; and second, foreign businesses which repatriated their profits to Europe naturally had less capital available to invest in further projects in India.[100]

Despite the existence of several macro-studies of these developments by economic historians, the role of technical experts in the industrialization process has received very little attention in the literature. Technology/technological know-how is seen as one of many factors in the development of industries, but is usually treated as a disembodied concept;[101] the literature presents no detailed descriptions

[99] Rothermund, *An Economic History of India*, p. 92; R. K. Ray, *Industrialization in India*, pp. 5, 271–2, 276; B. R. Tomlinson, *The Economy of Modern India, 1860–1970*, The New Cambridge History of India, III.3 (Cambridge: Cambridge University Press, 1996 [1993]), pp. 142–3, 145.

[100] R. K. Ray, *Industrialization in India*, pp. 244–5.

[101] B. R. Tomlinson remarked as early as 1981 that 'scholars studying the development of Indian industry in the first half of the twentieth century have paid too much attention to analyses based on generalized concepts of supply and demand constraints, technological innovation and adaptation, and factor endowments'. B. R. Tomlinson, 'Colonial Firms and the Decline of Colonialism in Eastern India 1914–47', *Modern Asian Studies*, vol. 15, no. 3 (1981): 455–86, here p. 485.

or systematic data regarding industrial engineers and technicians, their numbers, their qualifications, or their nationalities. Nevertheless, one significant theme can be discerned from remarks made in passing in the literature. This relates to the growing Indian-owned industries in our period. Most of them employed foreign technical experts and machinery in their early years, whether in the case of Sarupchand Hukumchand's engineering works, the Bombay textile industries that depended on plant and mechanical engineers from Lancashire, or the Indian shipping companies that could not initially obtain Indian marine engineers. It was particularly true of the specialized industries started in the 1940s, such as the automobile manufacturing concerns of the Birla and Walchand Groups. The literature also notes that efforts were made to train Indians to replace the foreign engineers in these industries.[102] Yet, how industrial enterprises went about this Indianization of their engineering staff is a question that has seldom been explored at any length.[103] It is this question, along with the more general one of assessing industrial engineers' contribution to the growth of large-scale industry that I address in detail in my study of the Tata Iron and Steel Company in Chapter 5.

***

Such were the key events and sentiments that influenced the Indianization of the public services in India. Three interlinked factors were at play. The first was the British policy of maintaining a sizeable proportion of British officers in the most responsible positions within the

---

[102] Iftikhar-ul-Awwal, *Industrial Development of Bengal*, p. 150; R. K. Ray, *Industrialization in India*, pp. 62, 105, 177–8.

[103] A very recent book by Ross Bassett explores a related theme, namely the experiences of Indians who studied at the Massachusetts Institute of Technology and returned to try and build industrial enterprises. Ross Bassett, *The Technological Indian* (Cambridge, MA, and London: Harvard University Press, 2016). While the term 'Indianization', as seen earlier, was primarily used in bureaucratic contexts, private industry at the time also used the term. See, for example, a 1929 diary entry by Edward Benthall of Bird & Co., Calcutta, which runs thus: 'The Bazaar talk is very badly against [Birla; presumably Indian industrialist G. D. Birla] and he has had [five] fires in four months at his mill—due to his Indianisation policy & refusal to employ Scotsmen.' Entry for 7 March 1929, Benthall's diary, Box 7, Benthall Papers, Centre of South Asia Studies (CSAS) Archives, University of Cambridge.

public services. The second was the growth of the nationalist movement, in both its moderate and radical forms. This led to constitutional reforms that, in amplifying the political voice of Indians in the
legislative councils, provided an effective forum in which to mount
demands for the Indianization of the services. The third factor was
that in the changing political and economic conditions in interwar
India, British aspirants became less enamoured of the Indian services. The history of the Islington and Lee Commissions should be
understood in the light of the state's attempt to balance these three
factors. Time and again, concessions to Indian aspirations were coupled with measures designed to make the services more attractive to
British recruits. At the same time, change was impeded by the colonial government's reluctance to allow rapid Indianization. The result
was that Indianization proceeded in small increments until the end
of the 1930s, when the Indian share of some of the major services
crossed 50 per cent. Whereas the existing historiography has sought
to understand these developments mainly in relation to the ICS, I am
concerned, in this book, with the effect of Indianization on engineers
and technical experts employed by the government in our period. The
later chapters in this book turn the spotlight on this group, focusing on the structure and composition of two important categories of
engineering service—public works and railways—that were deeply
affected by the negotiated process of Indianization and constitutional
reform.

Indianization, though, was not a phenomenon confined to government services. Indian industry, in terms of capital, was considerably
Indianized too, as various enterprises grew to maturity with the help
of protective tariffs in the 1920s and 1930s. As the scale of operations
increased, these could no longer depend solely on imported technical
experts. The resulting Indianization of expertise in large-scale industry is explored in this book through a case study of the Tata steel works
(Chapter 5). Industrialization also impacted the trajectory of the engineering profession, creating new employment opportunities and the
possibility of a new kind of identity for engineers. The interplay of
Indianization and industrialization with the professional identity of
engineers is the subject of the next chapter.

CHAPTER TWO

Putting Down Roots

Professional Institutions and the Growth of an Indian Identity among Engineers

I wish to draw the attention of our younger members to the field which India must shortly offer for employment to those who are capable of initiating and directing new industries. India, whose wealth has increased greatly during the war, does not need to rely in the future as in the past upon British capital, but she must rely for many years to come on British brains and expert knowledge.

—Edward Hopkinson, Presidential Address, Institution of
Mechanical Engineers (London), 1919[1]

Vast as India is, the problem of planning her economy is vaster.... Our Engineering conceptions have accordingly to be enlarged on a national scale, and what we need, both literally and in the wider sense, is in fact 'National Engineering'.

—Nawab Zain Yar Jung Bahadur, Presidential Address,
Institution of Engineers (India), 1945[2]

The changes in the Indian economy and the colonial state apparatus in the first half of the twentieth century were reflected, in varying

---

[1] 'Address by the President, Edward Hopkinson, Esq., D. Sc., M. P.', in *Proceedings of the Institution of Mechanical Engineers*, October–December 1919, pp. 631–58, here p. 649.

[2] Address printed in *The Journal of the Institution of Engineers (India)* (hereinafter cited as IEI *Journal*), vol. 26, no. 2, part 2 (December 1945): 37–42. Quoted text from p. 42.

degrees, in the experience of scientific and technical practitioners.[3] For engineers it was a particularly important phase, critical to the formation of their professional identity. This chapter traces the development of the Indian engineering profession as a whole in the period 1900–47.

One way of gaining historical insights into the collective development, aspirations, and identities of engineers is through the study of professional institutions—an approach that historians have used to study engineers in countries such as Britain, Germany, Canada, and the USA.[4] Here, I apply this approach to the engineering profession in India. As I will show, engineers in India organized themselves into groups to discuss technical aspects of their work. The nature and goals of their activities were influenced by various parameters including location, type of employer, and branch of engineering. They joined professional societies based in London, set up new ones in India that competed with each other, and used these institutions as a platform to define a pre-eminent role for themselves in the development of the Indian economy.

[3] See, for instance, David Arnold, *Science, Technology and Medicine in Colonial India*, The New Cambridge History of India, III.5 (Cambridge: Cambridge University Press, 2000), Chapter 5; Roger Jeffery, 'Recognizing India's Doctors: The Institutionalization of Medical Dependency, 1918–39', *Modern Asian Studies*, vol. 13, no. 2 (1979): 301–26. For other studies of groups of scientific/technical/medical practitioners in India covering this period or parts of it, see Ian Derbyshire, 'The Building of India's Railways: The Application of Western Technology in the Colonial Periphery 1850–1920' in *Technology and the Raj: Western Technology and Technical Transfers to India 1700–1947*, edited by Roy MacLeod and Deepak Kumar (New Delhi, Thousand Oaks, London: Sage Publications, 1995), pp. 177–215; Mark Harrison, *Public Health in British India: Anglo-Indian Preventive Medicine 1859–1914* (Cambridge, New York, and Melbourne: Cambridge University Press, 1994), chapter 1; Roger Jeffery, 'Allopathic Medicine in India: A Case of Deprofessionalisation?', *Economic and Political Weekly*, vol. 13, no. 3 (21 January 1978): 101–3, 105–13.

[4] R. A. Buchanan, *The Engineers: A History of the Engineering Profession in Britain, 1750–1914* (London: Jessica Kingsley, 1989); J. Rodney Millard, *The Master Spirit of the Age: Canadian Engineers and the Politics of Professionalism, 1887–1922* (Toronto: University of Toronto Press, 1988); Edwin T. Layton Jr, *The Revolt of the Engineers: Social Responsibility and the American Engineering Profession* (Baltimore: Johns Hopkins University Press, 1986); Kees Gispen, *New Profession, Old Order: Engineers and German Society, 1815–1914* (Cambridge: Cambridge University Press, 1989).

In the first part of this chapter, I demonstrate the importance of the London-based Institutions of Civil/Mechanical/Electrical Engineers for the predominantly expatriate engineers working in India in the nineteenth and early twentieth centuries. I then proceed to examine the history of the Institution of Engineers (India) or IEI, the first all-India professional institution for engineers. Established in Calcutta in 1920, the IEI aimed to unify engineers of all types (civil, mechanical, and electrical; government and industrial) under one umbrella. Its goals also included the Indianization of the profession and the promotion of industrial development. Drawing on a range of sources, including institutional membership data, journals, annual reports, and presidential addresses; government reports; and memoirs of practising engineers, I trace the circumstances leading to the IEI's founding. I then analyse its subsequent growth and development in the face of opposition from competing societies with different, more exclusive visions for the engineering profession. In particular, I argue that the IEI's growth to prominence marked the beginnings of a specifically Indian identity in the engineering profession, which had previously been oriented towards the imperial metropolis.

Thus, this chapter, through an analysis of the membership patterns, functions, activities, and agendas of the London institutions, local groupings in India, and the Institution of Engineers (India), charts and accounts for fundamental changes in the composition, priorities, and identity of the engineering profession in India over the first half of the twentieth century.

EXPATRIATE ENGINEERS AND METROPOLITAN INSTITUTIONS, 1858–1914

In the nineteenth and early twentieth centuries, the engineering profession in India was dominated by expatriate British engineers employed in public works, the military, and railways (some run by companies, others by the state).[5] Large-scale private industry—and

---

[5] Until 1905, the state-run railways formed a branch of the PWD; in that year a railway board was formed to govern all railways, and was itself responsible to a member of the Governor-General's Executive Council. Ian J. Kerr, *Engines of Change: The Railroads That Made India* (Westport, CT, and London: Praeger, 2007), p. 25 and pp. 75–9; G. Huddleston, *History of the East Indian Railway* (Calcutta: Thacker, Spink

the engineers associated with it—played a minor role in the Indian economy.[6] Thus, the main employers of engineers, with the exception of the company-run railways, were direct or indirect organs of the colonial state, interlinked in function, personnel, and organizational structure. They recruited primarily in Britain, accounting for the largely British composition of the profession (Figure 2.1).

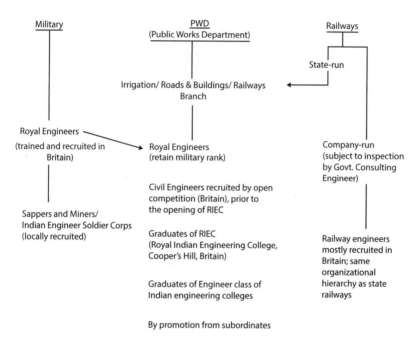

Figure 2.1 Major employers of engineers in India, c. 1850–1905

Source: Compiled by the author from various sources.

and Co., 1906), chapter 1; *East India (Railway Committee, 1920–21). Report of the Committee Appointed by the Secretary of State for India to Enquire into the Administration and Working of Indian Railways* (Cmd. 1512, London: HMSO, 1921), chapter IV, pp. 36–7. The last of these sources is hereinafter cited as *Acworth Committee Report*.

[6] Considering industrial income alone in 1900, large-scale industries' contribution to it was only 15 per cent. Tirthankar Roy, *The Economic History of India 1857–1947*, 2nd edn (New Delhi: Oxford University Press, 2006), p. 224.

As the numbers of British engineers in India grew,[7] some began to meet at local mechanics' institutes to discuss their work experiences and organize lectures on topics of common interest. These mechanics' institutes appear to have been modelled on the contemporary British ones.[8] An example is the Sassoon Mechanics' Institute (est. 1863), which grew out of an association of mechanics from the Royal Mint and Government Dockyard of Bombay, with a benefaction from the banker David Sassoon of that city.[9] Mechanics' institutes, however, were not professional institutions, and no large-scale society of engineers was established in India in the nineteenth century.[10] Instead, many engineers became members of the professional institutions then gaining ground in Britain.[11] Prominent among these were the London-based trio of the Institution of Civil Engineers (ICE, est. 1818), the Institution of Mechanical Engineers (IMechE, est. 1847), and the Institution of Electrical Engineers (IEE; founded in 1871 as the Society of Telegraph Engineers).[12]

The three London institutions were similar in structure and function. Their primary aim was to enable practising engineers to discuss papers on matters of common technical interest. They also performed what Angus Buchanan has called the 'informal licensing function' by obtaining Royal Charters or legal incorporation. Although engineers did not legally require a licence to operate, corporate status indicated that the state endorsed an institution, and thereby its members.[13] To

---

[7] There were 518 PWD engineers in India in the year 1861. Brendan P. Cuddy, 'The Royal Indian Engineering College, Cooper's Hill, (1871–1906): A Case Study of State Involvement in Professional Civil Engineering Education' (PhD thesis, London University, 1980), p. 62, table VIII.

[8] Derbyshire, 'The Building of India's Railways', p. 182.

[9] See 'David Sassoon Library and Reading Room: History', available at http://www.davidsassoonlibrary.com/history.html (accessed 2 July 2012).

[10] A record exists of an Indian Engineers' Association set up in Calcutta in 1871, mainly by Indians, but it is not known if its membership ever extended beyond Bengal, or how long the association survived. See *Transactions of the Indian Engineers' Association for the Half Year Ending June 1871*, vol. I (Calcutta: Printed by H. C. Ganguly & Co, 1872).

[11] Derbyshire, 'The Building of India's Railways', pp. 182–3.

[12] On the origins of the London institutions, see R. A. Buchanan, 'Institutional Proliferation in the British Engineering Profession, 1847–1914', *The Economic History Review*, new series, vol. 38, no. 1 (February 1985): 42–60.

[13] Buchanan, 'Institutional Proliferation'. The quoted phrase is from p. 59.

ensure that membership was equated with high professional standards, the institutions established strict requirements—including practical experience and entrance examinations—for aspiring members. Buchanan writes that they also probably performed a regulatory function by '[acting] as a constructive restraint on professional irregularities' and a business function by matching projects to practitioners, 'making expertise available where it was needed'.[14]

The ICE, IMechE, and IEE were international in scope. Beyond Britain, they had members in India, in other parts of the British Empire, and sometimes outside it, for example in Japan and China.[15] These expatriate engineers considered themselves part of an imperial profession with its base in Britain. They often sent accounts of their work to the London institutions, and sometimes attended meetings when they went home on leave. Nineteenth-century issues of the *Minutes of the Proceedings of the Institution of Civil Engineers* contain several instances of papers on engineering projects in India, Egypt, Australia, and South Africa.[16] Members received comments and questions in person from colleagues in Britain when the paper was read, and by correspondence from colleagues elsewhere. This enabled the comparison of notes on techniques adopted in different parts of the world.[17] Over time, the process of communication between colonial members and the metropolitan councils was systematized. Provision was made for 'colonial representatives' on the ICE's Council—in India's case, starting from 1898.[18] The ICE also set up Advisory Committees in South Africa and several Australian provinces (at various times

[14] Buchanan, 'Institutional Proliferation'. Quoted text from p. 57.

[15] (Topographical) lists of members of the three institutions, various years.

[16] Derbyshire, 'The Building of India's Railways', pp. 178, 183, 206n20; titles of papers presented on projects in various countries can be seen in the *Minutes of the Proceedings of the Institution of Civil Engineers* of various years (via www.icevirtuallibrary. com).

[17] See, for example, Bradford Leslie, 'The Erection of the "Jubilee" Bridge Carrying the East Indian Railway across the River Hooghly at Hooghly', *Minutes of the Proceedings of the Institution of Civil Engineers*, vol. 92, no. 1888: 73–96; also the discussion (*Minutes of the Proceedings*, pp. 97–128) and correspondence on the paper (*Minutes of the Proceedings*, pp. 129–41). See particularly the contributions of J. H. Cunningham, R. E. Johnston, and Leslie himself in reply, in the correspondence.

[18] Memorandum on 'Election of Colonial Representatives on the Council', ICE Archives, London: 185/01.

between 1890 and 1907), in New Zealand (1906), in Canada (1911), and in India (1912).[19] The function of these committees appears to have been to advise the Council in London on matters such as admission of members in the concerned overseas location, and their transfer to higher grades of membership.[20] In 1915, the IMechE followed suit and initiated the process of setting up advisory committees in India and South Africa.[21]

The India-based membership of the London institutions reflected the predominantly British composition of the engineering community. In 1901, Europeans formed approximately 94 per cent of ICE members residing in India. The corresponding number for the IMechE was also 94 per cent (in 1901) and for the IEE 87 per cent (in 1902). The dominant institution, numerically, was the ICE, which in 1901 had 538 members resident in India, while the IMechE had 124. The corresponding figure for the IEE was 99 members in 1902 (see Tables 2.1, 2.2, and 2.3 later in this chapter). The comparatively high numbers for the ICE are a reflection of the overwhelming importance of civil engineering over other branches in India at this time, when the construction of public works and railway lines constituted the bulk of engineering activity—mechanized industries, with the exception of the cotton and jute mills of Bombay and Calcutta, being virtually non-existent.[22] Indeed, many British engineers who joined the Public Works Department (PWD) in the early twentieth century would already have been members of the ICE before their first posting, as

---

[19] 'The Institution of Civil Engineers: Advisory Committees', ICE Archives, London: 185/01.

[20] According to Garth Watson, these were the functions of the advisory committees set up in yet other countries in the late 1940s. Garth Watson, *The Civils: The Story of the Institution of Civil Engineers* (London: Thomas Telford, 1988), pp. 235–6.

[21] IMechE *Proceedings*, January–May 1916, p. 12 (mentioned in the Annual Report for 1915). For a detailed analysis of the ICE's membership in the wider British Empire (particularly in Africa) and the power relations between these engineers and their metropolitan counterparts, see Casper Andersen, *British Engineers and Africa, 1875–1914* (London: Pickering & Chatto, 2011 [Kindle edition consulted]), chapter 4.

[22] On the state of large-scale industries before World War I, see T. Roy, *The Economic History of India*, pp. 227–8; R. K. Ray, *Industrialization in India: Growth and Conflict in the Private Corporate Sector 1914–47* (New Delhi: Oxford University Press, 1982 [1979]), p. 4.

Associate Membership of the ICE was one of the qualifications that made one eligible to apply for a PWD position.[23]

Within the broad identity of engineering in the British Empire, those working in India also began to see themselves, by the final decades of the nineteenth century, as having a particular expertise in engineering in the subcontinent. As Ian Derbyshire has shown, most of them spent decades, sometimes their whole working lives, in India. In many areas of railway engineering, they evolved a 'technologically syncretic' style of engineering that included elements of traditional Indian practices, often more labour-intensive than the prevailing Western norm.[24] In keeping with Derbyshire's argument about syncretism, examples may be found of India-based engineers referring to conditions peculiar to India in explaining certain features of their work to their metropolitan colleagues.[25]

Further, a shared experience of training undergone at colleges oriented towards Indian government services added to the sense of community among engineers working in different parts of India. These colleges (which are discussed further in Chapter 3) included the Royal Indian Engineering College (RIEC) at Cooper's Hill near London, set up in 1871 specifically to train engineers for the PWD, and the four Indian engineering colleges established in the nineteenth century at Roorkee, Madras, Poona, and Sibpur (near Calcutta). While the full-time students of the Indian colleges were permanent residents of India, it appears that fresh engineers from Britain sometimes spent time at one of them as part of their acclimatization to India.[26]

---

[23] 'Indian Public Works Department. Regulations as to Appointments of Assistant Engineers, 1910' in the *India Office List* for 1910, p. 222.

[24] Derbyshire, 'The Building of India's Railways'. The quoted phrase is from p. 183.

[25] See, for example, Joseph F. Strong, 'On the Apparatus Used for Sinking Piers for Iron Railway Bridges in India', *Proceedings of the Institution of Mechanical Engineers*, vol. 14, no. 1 (1863): 16–33 (including discussion). Strong referred in the discussion to the use of 'pounded brick instead of sand ... for making the mortar' for the cylinders used in the bridge piers. He said that pounded brick 'was used everywhere throughout India in preference to sand, because the sand was too fine' (Strong, 'On the Apparatus Used for Sinking Piers', p. 28.) See also Bradford Leslie in the discussion and correspondence on his paper 'The Erection of the "Jubilee" Bridge'.

[26] Derbyshire, 'The Building of India's Railways', pp. 181–2; see also Cuddy, 'Royal Indian Engineering College', chapter 2.

There was also some exchange of news and technical ideas with the founding of journals published in India, such as *The Indian and Eastern Engineer* and *Indian Engineering*.[27] The former was begun in 1858 by C. C. Adley,[28] a senior telegraph engineer in the company-run East Indian Railway. In its early years the journal bore the title *The Engineer's Journal and Railway, Public Works, and Mining Gazette, of India and the Colonies*; it was published from Calcutta and circulated across India.[29] The other journal, *Indian Engineering*, was started in 1887 as a weekly, under the editorship of Pat. Doyle, C. E. (Civil Engineer).[30] This journal contained a wide variety of articles dealing with PWD engineering, railway work, and mill machinery among other topics,[31] while the editor stressed that contributions should relate to 'actual Indian practice'. The journal invoked a collective identity for engineers in India, and aimed to provide a 'means of intercommunication' between engineers belonging to '[t]he Profession throughout the country [which] is fast increasing in numbers and influence'.[32]

Local Organization

As this incipient colonial identity grew, engineers in India began to form local chapters of the London institutions.[33] Around the 1910s,

[27] Derbyshire, 'The Building of India's Railways', p. 182.

[28] Derbyshire, 'The Building of India's Railways', p. 182.

[29] *The Engineer's Journal and Railway, Public Works, and Mining Gazette, of India and the Colonies*, vol. IV (Calcutta: Savielle and Cranenburgh, Printers, Bengal Printing Company Limited, 1861). As of 1861, C. C. Adley was Superintendent, East Indian Railway Electric Telegraphs. See the 'Address' to him from Radhamadhub Chuckerbutty and others, *Engineer's Journal*, vol. 4 (2 December 1861): 219.

[30] 'Ourselves', *Indian Engineering*, 1 January 1887, pp. 1–2. The reference for the full volume is *Indian Engineering: An Illustrated Weekly Journal*, edited by Pat. Doyle, C.E., vol.1, January–June 1887 (Calcutta: Printed by C. J. A. Pritchard at the Star Press, 19 Lall Bazar Street, 1887).

[31] 'General Index', *Indian Engineering*, vol. I, pp. i–vi.

[32] 'Ourselves', *Indian Engineering*, 1 January 1887, pp. 1–2.

[33] The historian Vanessa Caru's ongoing research into the engineers of the Bombay PWD has revealed the existence of an Association of Indian College Engineers from 1910 (Vanessa Caru, 'The creation of a "corps d'état"? Indian and European engineers of the Bombay Public Works Department [1860's–1940's]', presented at the midterm workshop of the project ENGIND: Engineers and Society in Colonial and Postcolonial India [under the French National Research Agency], Delhi, 11 January 2016). However,

the ICE and IEE tried to set up local sections;[34] only the latter suc-
ceeded, establishing a Calcutta centre, though it has not been possible
to ascertain how active it was.[35] Members of the IMechE appear to
have had the most success in this direction. In 1907 IMechE members in and around Calcutta started to organize meetings to discuss
papers. In January 1909 they formed, with the approval of the London
Council, a Calcutta and District Section of the IMechE. Here, members could meet face to face, present original papers, and discuss
papers that had been read in London; some of the Calcutta papers and
discussions were in turn excerpted in the IMechE's *Proceedings*.[36] By
1915–16 the Section had eighty-eight members and was organizing
visits to engineering works in the Calcutta region.[37] Meetings were
then suspended on account of the Great War.[38] According to a government report that commented on the matter in passing, the Section
only had limited success as it '[had] not sufficient prestige to attract
many members from other parts of India'.[39]

Even as the Calcutta Section suspended its activities, its Chairman
began to receive letters from 'influential members' of the IMechE in

---

I have been unable to ascertain any further details about the Association's scope or
importance, and have not come across any other references to it in the primary sources
I have seen. I thank Dr Caru for permission to cite her work in progress.

[34] 'Address by the President Edward Hopkinson, Esq., D.Sc., M. P.', October 1919,
in IMechE *Proceedings*, October–December 1919, pp. 631–58, here p. 655.

[35] IEE List of Members (as of 1 October 1919), p. 5. That the ICE did not succeed
in setting up a local centre at this time is inferred from Hopkinson's Address (cited
earlier), and confirmed by the ICE's official correspondence of 1926, considering
afresh the creation of a local centre in India. ICE Archives, London: 185/03.

[36] See Annual Report for 1909, IMechE *Proceedings*, February 1910, pp. 343–83,
here pp. 346–7, 360. For an example of a paper presented first at Calcutta and
then published in the *Proceedings*, see J. M. Christie, 'The Manufacture of Oxygen
and Its Use for Welding and Metal-Cutting', IMechE *Proceedings*, December 1916,
pp. 889–94.

[37] Annual Report for 1916, IMechE *Proceedings*, January–May 1917, pp. 1–40, here
p. 14.

[38] Annual Report for 1917, IMechE *Proceedings*, January–May 1918, pp. 1–41, here
p. 11.

[39] Appendix H: 'Scientific and Technical Societies', *East India (Industrial Commission). Report of the Indian Industrial Commission 1916–18* (Cmd. 51, London: HMSO,
1919), pp. 385–7, here p. 386. This source is hereinafter cited as IIC Appendix H.

India proposing possible alternative arrangements. In particular, they asked for a fully fledged Indian branch with a secretary, clerical staff, and its own lecture hall and reading room. This correspondence was forwarded to London. The Council, partly because of a lack of funds, declined to implement the suggestion. Instead, it suggested the amalgamation of local sections of the London institutions in order to share the cost of providing technical libraries and other facilities. Apparently, this did not materialize.[40] It is not clear why members asked for an all-India branch as opposed to reviving the Calcutta Section (which appears to have lapsed permanently at the end of the war).[41] The reference to a lecture hall, reading room, and staff suggests that what they wanted was a more permanent and more prestigious Indian base than the existing Calcutta Section.

Meanwhile, PWD engineers in India were organizing themselves on a different basis. The process began in Punjab, which, with its network of irrigation canals, had more public works engineers than any other province in the country.[42] In 1912, ninety PWD engineers in that province, of whom sixteen were Indians, came together to form the Punjab PWD Congress (the name was changed to Punjab Engineering Congress in 1916).[43] The Congress aimed 'to promote the well-being of the Department by affording members an opportunity of meeting annually to discuss subjects of professional or Departmental interest and for social intercourse'.[44] The founding Secretary, who continued in that post for several years, was W. S. Dorman, MICE[45]

[40] Annual Report for 1918, IMechE *Proceedings*, January–May 1919, pp. 1–42, here pp. 10–11.

[41] The *Proceedings* of later years do not mention the Calcutta and District Section (as indicated by a digital search on http://pme.sagepub.com, accessed on 3/4 July 2012).

[42] See, for example, the listing of PWD engineers by province in the *India Office List* for 1914.

[43] 'History: Pakistan Engineering Congress, An Overview of 94 years', http://pecongress.org.pk/history.php, accessed 3 July 2012. The article is on the website of the successor institution of the Punjab Engineering Congress in Lahore.

[44] 'Rules', back matter (page numbers not contiguous with main text), *Minutes of Proceedings of the Punjab P.W.D. Congress*, vol. I (Lahore: Civil and Military Gazette Press, 1913) [British Library Shelfmark: ST 1811].

[45] MICE: Member of the Institution of Civil Engineers; AMICE: Associate Member of the Institution of Civil Engineers.

(then an Executive Engineer in the Punjab PWD) and the first President was Sir H. P. Burt, AMICE (a senior railway official).[46] Gazetted (officer rank) engineers in the PWD and state-run railways could become members of the Congress. Municipal and district engineers holding recognized diplomas were permitted to join as Honorary Members, as were Royal Engineers from 1913. Residence in Punjab was not a requirement, and the annual subscription was Rs 10.[47]

The Congress was an annual meeting where papers were presented and discussed. Between 1913 and 1920, an average of ten engineering papers a year were presented at the Congress, whose membership at the end of this period was 220.[48] The Congress appears not to have had any regulatory or licensing functions. Nor did it have buildings of its own—Committee meetings took place at various venues in Lahore, including the Town Hall and the PWD Secretariat. It was primarily a forum where PWD engineers could get to know a wide spectrum of their colleagues from the Punjab and elsewhere. Prestige was a factor: individual members delivering papers stood to win recognition not just from their immediate superiors but from a large body of officers, and could sometimes showcase their ideas or achievements to high government officials. The Lieutenant Governor of Punjab was invited to some of the annual sessions, as in 1920, when he described two papers from the previous year as having influenced him and his government in inaugurating specific engineering schemes.[49]

But the founders of the Punjab Congress had a larger ambition. They wanted similar PWD Congresses to be formed in other provinces, which could then be federated into an All-India PWD Congress, which would seek affiliation with the ICE in London. W. S. Dorman, the first Secretary, reported that in 1912–13 a Committee entrusted

---

[46] Listed in *Minutes of Proceedings of the Punjab P.W.D. Congress*, vol. I (1913). On Dorman's long innings as Secretary, see A. S. Montgomery, Presidential Address, April 1920, in *Minutes of Proceedings of the Punjab Engineering Congress*, vol. VIII (Lahore: Civil and Military Gazette Press, 1920): i–vii, here pp. i–ii. On Dorman's career, see Record of Services in *India Office List* for 1929. H. P. Burt became President of the Indian Railway Board in 1914–15 (*Acworth Committee Report*, p. 3).

[47] 'History: Pakistan Engineering Congress'.

[48] Montgomery, Presidential Address, p. ii; 'History: Pakistan Engineering Congress'.

[49] 'History: Pakistan Engineering Congress'.

with promoting the all-India plan had sent out a letter to Chief Engineers across India, inviting to its next meeting 'any engineers desirous of taking the opportunity of seeing how the movement in the Punjab has been progressing'.[50] The plans came partly to fruition. By 1915, PWD engineers in Bombay and Burma had formed Congresses in their respective provinces. Although there remains the possibility that this was an independent development, the existence of a measure of cooperation by 1920 is indicated by the listing of papers read at these societies in the *Minutes* of the Punjab Congress.[51]

Therefore, although the Punjab PWD Congress was not a professional institution in the sense of the ICE or IMechE, its development is revealing of the kind of identity its founders wanted to build: an elite identity as engineer officers of the colonial government. By restricting itself to gazetted officers, the Punjab Congress limited its potential membership: in effect, it excluded all engineers who were not employed by the government, for example, engineers in company-run railways or private industries.[52] The 'profession' the Congress represented was thus a very specific one, which mapped directly onto the government engineering services across India. One member even hoped that those who belonged to the envisioned All-India Congress would eventually receive from the government an honorific such as CIES (Congress of the Indian Engineering Services), as a 'hall-mark of distinction'.[53] The idea of a post-nominal analogous to 'ICS'—which members of the prestigious Indian Civil Service bore— suggests an aspiration to a readily identifiable marker of status as a

[50] Dorman, Annual Report and R. E. Purves, 'P. W. D. Congress. Note on the Best Means to Be Adopted for Extending Its Influence', back matter, p. 89 (page numbers not contiguous between main section and back matter) in *Minutes of Proceedings of the Punjab P.W.D. Congress*, vol. I.

[51] *Minutes of Proceedings of the Punjab Engineering Congress*, vol. VIII, pp. 135–42.

[52] Edward Hopkinson, in his 1919 presidential address to the IMechE, observed that '[t]he Members of the Public Works Department Congresses have realized that exclusion of all except Government Engineers, which was their policy until recently, was a mistake, and are taking steps to widen their constitutions' ('Address by the President Edward Hopkinson', p. 655). The Punjab Congress eventually made their membership requirements less specific. See 'History: Pakistan Engineering Congress' on Ganga Ram's presidential year (1923/24).

[53] Purves, 'P. W. D. Congress', p. 92.

colonial government administrator.[54] The Congress's goal of being affiliated with the ICE also suggests that for the mainly British pioneers of the Punjab Congress, their Indian identity remained only a part of their larger identity as imperial engineers.

The foregoing discussion has highlighted the engineering profession's predominantly European composition, its bureaucratic character, and the metropolitan/imperial identity of its members in the long nineteenth century. These features, however, began to change in the interwar period.

WAR, GOVERNMENT POLICY, AND THE FORMATION OF AN INDIAN INSTITUTION (c. 1900–20)

As shown in Chapter 1, World War I catalysed significant constitutional reforms in India and caused fundamental changes in the colonial government's industrial and fiscal policies. As a result, the structure of government services, the nature of the economy, and the role of Indian professionals in both were transformed considerably in the interwar years. Indianization was set in train by the appointment of Royal Commissions, while a number of industries received economic protection, enabling the expansion of large-scale industrial enterprises in the 1920s and 1930s.

These changes had a marked effect on the composition of the engineering profession in India, which expanded, diversified, and was considerably Indianized. While the details of these changes are examined in the next three chapters, an analysis of membership trends of the London-based professional institutions illustrates the overall trajectory. The following tables (Tables 2.1, 2.2, and 2.3) show these statistics for the first four decades of the twentieth century.

Some patterns are immediately visible. First, the overall membership of the three institutions in India showed a substantial increase over this period, indicating an expanding engineering profession. Second, this increase was much more dramatic for the IMechE and the IEE than for the ICE (whose membership started at a much

---

[54] As it happened, engineers in the top PWD ranks were able to use the letters ISE, for Indian Service of Engineers, after 1920; but this had nothing to do with membership of the PWD Congresses. On the ISE, see Chapter 3.

Table 2.1 Membership trends: Institution of Civil Engineers (ICE)

Year	ICE			
	No. of India-resident members	As a percentage of worldwide membership	Number of 'native' names among India-resident members	'Native' names as a percentage of India-resident members
1901	538	7.32	30	5.58
1940	643	4.98	213	33.13

Sources: Tables 2.1, 2.2 and 2.3 are based on the Lists of Members of the ICE, IMechE and IEE respectively (consulted at the archives of the ICE, IMechE and IET respectively, all in London). The figures are only indicative, for a variety of reasons: (*a*) where topographical listings are not made, the number of India-resident members is estimated by counting names with identifiably Indian addresses; (*b*) the count for 'native' names may not account for some Anglo-Indian names that are indistinguishable from European names; (*c*) occasional anomalies occur, such as the appearance of names without addresses—in which case they are not taken into account. Further, Burma (Myanmar) and Ceylon (Sri Lanka) are sometimes listed along with India and sometimes separately. For consistency, they have been included in all totals given here. Finally, wherever possible, the latest list for a particular year has been used.

Note: The number of ICE members in India appears not to have varied much over this period. According to a contemporary, the ICE had about 600 members in India in 1919. See 'Address by the President Edward Hopkinson, Esq., D.Sc., M.P.', October 1919, in IMechE *Proceedings*, October–December 1919, pp. 631–58, here p. 654.

Table 2.2 Membership trends: Institution of Mechanical Engineers (IMechE)

Year	IMechE			
	No. of India-resident members	As a percentage of worldwide membership	Number of 'native' names among India-resident members	'Native' names as a percentage of India-resident members
1901	124	3.87	7	5.65
1919	360	5.45	25	6.94
1930	620	5.77	90	14.52
1940	691	5.05	214	30.97

higher number but did not increase as rapidly), indicating that the major growth area in Indian engineering after World War I was in private industry, which accounted for most electrical and mechanical engineers. Third, the percentage of 'native' members among the

Table 2.3   Membership trends: Institution of Electrical Engineers (IEE)

| Year | IEE | | | |
	No. of India-resident members	As a percentage of worldwide membership	Number of 'native' names among India-resident members	'Native' names as a percentage of India-resident members
1902	99	2.25	13	13.13
1919	299	3.84	75	25.08
1930	771	5.47	372	48.25
1939	1,445	7.57	993	68.72

India-resident members of the three institutions increased dramatically, suggesting an increasing Indianization of the profession. Thus, the engineering profession reflected the interwar trends of industrialization and Indianization.

In keeping with these trends, there occurred another important development in 1920: the establishment of a new professional society on Indian soil, the Calcutta-based Institution of Engineers (India). This institution owed its existence in large part to the report of the Indian Industrial Commission (IIC) (1916–18). In proposing that the state should abandon its *laissez faire* policy and actively encourage industrialization, the IIC's report highlighted the government's need for 'reliable scientific and technical advice'.[55] To provide such advice, the Commission proposed not only to reorganize the government's scientific services, but also to set up professional societies that would include privately employed technical experts in addition to government officers.[56] They envisioned these as 'Indian institutes, societies and associations analogous to the Institution of Civil Engineers, the Chemical Society, and the British Association for the Advancement of

[55] 'The Indian Industrial Commission: Its Report Summarised', in *East India (Industrial Commission)*. *Report of the Indian Industrial Commission 1916–18* (Cmd. 51, London: HMSO, 1919), p. 1.

[56] 'The Indian Industrial Commission: Its Report Summarised'; also Indian Industrial Commission 1916–18, *Report* (Calcutta: Superintendent Government Printing, India, 1918), point 42, p. 277. The last of these sources is hereinafter cited as IIC *Report*.

Science'.[57] One such society would be 'an Indian Institution of Engineers embracing all the branches of engineering practised in India'. This would guide the government in its project of industrialization, advising it on regulatory matters like the certification of boiler attendants, laws relating to engineering, and licences for mineral resources and water power. It would also advise on engineering education, and help in 'establishing a standard of professional conduct and efficiency' among engineers.

As the IIC envisioned it, the new institution would be significant in two ways. First, it would embrace private as well as public engineers. This was important, as industrialization would mean more engineers employed in private industry. Second, along with the other proposed professional societies, it would help meet 'the growing needs of Indians' by allowing them to participate in professional meetings and discussions. Although they could join the English and American professional societies, Indian engineers did not gain much from this 'beyond the prestige attaching to membership and the periodic receipt of copies of publications', as they did not ordinarily have opportunities to travel to Britain or the USA. The new Institution of Engineers, in contrast, would be located in India. Furthermore, the IIC wanted it to follow the practice of the IMechE in Britain by setting up local centres in different parts of India, thus allowing greater numbers of engineers to attend meetings, and to present and discuss papers.[58] However, as discussed later, this did not mean that the institution would be solely for Indians—only that they would be as welcome as Europeans.

In India Sir Thomas Holland, President of the IIC, led the movement for the new institution. Holland had previously played leading roles in other professional institutions. He was a co-founder (1906) and the first President of the Mining and Geological Institute of India; President of the Manchester Geological and Mining Society (1913); and President of the Institution of Mining Engineers, London (1915–16).[59] As Director

[57] IIC Appendix H, p. 385. The two subsequent quotes in the paragraph are also from this Appendix.

[58] IIC Appendix H.

[59] L. L. Fermor, 'Thomas Henry Holland. 1868–1947', *Obituary Notices of Fellows of the Royal Society*, vol. 6, no. 17 (November, 1948): 83–114, here p. 90.

of the Geological Survey of India (1903–c. 1910),[60] Holland had proved himself a keen administrator and reformer, and was instrumental in the Survey's attainment of 'a position of prestige in India, both with Government and with the public'.[61] In 1917 he became President of the Munitions Board,[62] and was thus a central figure in India's contribution to the British war effort. The proposed Indian Institution of Engineers could scarcely have had a more influential champion.[63]

The proposal coincided with the demands for all-India branches (as outlined earlier) by the India-resident members of some of the London engineering institutions. The 'home' institutions—in particular, the IMechE and the IEE—could now meet these demands by asking their members in India to join the new Indian institution, over which they initially harboured hopes of exerting control from London. As the annual report of the IMechE for 1918 put it,

> The Council have ... expressed their willingness, should the members resident in India so desire, to consider [a] proposal which they are aware is under discussion in India for the establishment of an Indian Institution of Engineering, incorporated by charter or otherwise, so as to form a corporate body of the members of the Institutions of Mechanical and Electrical Engineers in India, in close association with and affiliated to the Home Institutions, and receiving their financial support. The Council, however, consider that full membership of such an Institution, that is, membership with full privileges and voting power, should be dependent on membership of one of the Home Institutions, though not thereby precluding admission to other grades of membership with limited privileges, such as Associate

---

[60] Roy M. MacLeod, 'Holland, Sir Thomas Henry (1868–1947)', *Oxford Dictionary of National Biography*, Oxford University Press, 2004 (available at http://www.oxforddnb.com/view/article/33945, accessed 11 June 2012).

[61] Fermor, 'Thomas Henry Holland', p. 85.

[62] MacLeod, 'Holland, Sir Thomas Henry (1868–1947)'.

[63] Holland's importance in this matter is indicated in several ways. He was Chairman of the Industrial Commission, which recommended the setting up of the Institution; he presided over a meeting in Calcutta where the decision to form the new Indian institution was taken (see 'Address by the President Edward Hopkinson', here p. 655); IIC Appendix H appears to draw on Holland's experience, mentioning the Mining and Geological Institute of India as an example of previous successful institutions in India.

and Graduate, subject to qualifications similar to those now required by the Home Institutions.[64]

The interest in maintaining some influence over the Indian institution remained, but the idea of direct metropolitan control was soon dropped—possibly because it met with resistance in India. That some negotiation occurred is likely: the new President (1919) of the IMechE, Edward Hopkinson, would have known personally the proponents of the Indian institution, as he had himself been a member of the IIC before resigning when he was advised, on medical grounds, against travelling to India in November 1917.[65] As Hopkinson explained in his presidential address of October 1919,

> It is [now] generally conceded both here and in India that it is better that the Indian Society, knowing the wants of its own members and the local conditions, should devise its own constitution independently. At the same time, it is, I think, incumbent upon our Institution to use its influence in maintaining a high qualification for membership in the Indian Society and to collaborate with it with a view to obtaining the benefits for our own members in India, which such a Society can provide.[66]

In other words, the London institutions found it expedient to support the proposal for a new institution in India, but were anxious that such association should not be construed as lowering the metropolitan standards of the profession. Stringent entry requirements to the Indian institution would ensure that the IMechE's members in India would not, in joining or working with the new institution, have to associate with a local profession of a lower standard.

Nevertheless, the endorsement of the London institutions continued, and with the influential figure of Thomas Holland in India to shepherd it through, plans for the new Indian institution progressed swiftly. Prominent engineers met in various parts of the country to discuss the proposal. Holland organized an initial meeting in Calcutta in 1919; an organizing committee was formed; a constitution

[64] IMechE *Proceedings*, January–May 1919, pp. 11–12.

[65] 'Engineering Heritage: Past Presidents', available at http://heritage.imeche.org/Biographies/pastpresidents (accessed 14 April 2012); IIC *Report*, pp. xv–xvi.

[66] 'Address by the President Edward Hopkinson', p. 656.

was drafted and circulated among engineers across India. In September 1920 the Institution of Engineers (India) or IEI was formally registered under the Indian Companies Act.[67] It started functioning from Calcutta soon after, whereupon '[t]he goodwill of the older Institutions' was 'promptly and handsomely expressed'.[68]

The London institutions, in supporting the new institution, recognized the fact that it was not in direct competition with them. Joining the IEI did not mean that engineers in India would cease to become members of the ICE, IMechE, or IEE: as Tables 2.1, 2.2, and 2.3 show, large numbers of both Britons and Indians continued to be members of these institutions in the decades after the IEI was founded. Yet, for a community of engineers that had earlier seen itself as part of an Empire-wide profession, the IEI signified the possibility of another, more specific, identity. Its formation was part of a wider trend: practitioners of science, medicine, and technology (STM) had begun to organize themselves in pan-Indian societies. David Arnold has described the growth of an 'Indian scientific community' from 1890 onwards, which entered a defining phase with the formation of the Indian Science Congress in 1914. The Science Congress enabled the participation of university-based scientists (and more Indians) alongside members of the (European-dominated) government scientific services in an annual discussion of papers.[69] An Indian Medical Association was formed in 1928 by doctors opposed to metropolitan control of the profession and its domination by the Indian Medical Service, although it was some years before it was able to bring about any changes in the situation.[70] All these developments were related directly or indirectly to the twentieth-century constitutional reforms that paved the way for greater Indianization and provincial autonomy. The IIC was conscious of these broader developments in STM: in recommending the creation of an Indian

[67] H. Nandy (ed.), *IEI Marches On* (Kolkata: Cdr. A. K. Poothia for Institution of Engineers [India], 2002 [1996]), pp. 5–6.

[68] Hugh W. Brady, 'The Origin, Character and Prospects of the Institution', IEI *Journal*, vol. 1 (September 1921): 30–7, here p. 37.

[69] Arnold, *Science, Technology and Medicine in Colonial India*, chapter 5.

[70] Jeffery, 'Recognizing India's Doctors'. According to the IMA's website, many of its early proponents were 'active in the Indian national Congress'. 'IMA in Retrospect', available at http://www.ima-india.org/IMA_history.html (accessed 12 April 2012).

institution for engineers, the IIC had referred approvingly to the early
success of the Indian Science Congress.[71]

THE IEI AND THE GROWTH OF AN INDIAN
ENGINEERING PROFESSION, 1920–47

The establishment of the IEI and its interwar activities offer important insights into the emergent Indian professional identity of engineers, and their relationship with the state. Throughout this period,
the IEI's primary source of legitimacy was its endorsement by the
colonial state. The Institution was an incorporated body from the
beginning (registered, as we saw above, under the Indian Companies
Act). In 1935 it went a step further, obtaining a Royal Charter.[72]

The text of the 1935 Royal Charter—which formalized the IEI's
existing goals[73]—gives an accurate idea of the envisaged functions
of the IEI throughout the period 1920–47. Broadly speaking, the IEI
was similar to the London institutions in form and purpose. It aimed
to promote technical knowledge through paper discussions and publications; to play a role in engineering education through classes and
examinations for working persons; to provide technical advice to the
government on behalf of the engineering profession; and to act as a
governing body to regulate the conduct of its members and 'promote
efficiency and just and honourable dealing and ... suppress malprac-
tice in engineering'.[74] Despite these similarities, though, three particular objectives of the IEI are worth stressing as being somewhat
different from those of the London Institutions. First, the Institution
was to encompass all branches of engineering.[75] Second, it sought to

---

[71] IIC Appendix H, p. 386.

[72] Nandy, *IEI Marches On*, p. 11.

[73] As will be seen in this section, the IEI's activities were essentially marked by
continuity before and after the award of the Charter (1935).

[74] Annexure IV: 'Royal Charter', Nandy, *IEI Marches On*, pp. 78–88.

[75] According to the IEI's Charter, the overall objective of the Institution was
'[t]o promote and advance the science, practice and business of Engineering in all its
branches ... in India' (Nandy, *IEI Marches On*, pp. 78–88).The origins of the decision to
cover all branches of engineering lay at least partly in pragmatic considerations. As the
IIC had observed, there were not yet sufficient numbers of engineers in India to form
several specialized institutions of engineers (IIC Appendix H, p. 386). In any event, the

promote Indian industrialization, reflecting its genesis in the recommendations of the IIC.[76] Third, it was to encourage Indianization in the engineering profession—again reflecting the Industrial Commission's view, quoted earlier, that professional societies were needed in India to cater to the 'the growing needs of Indians'. Later, I will show how the Institution attempted to fulfil these manifold functions, and relate this to the emergence of an Indian professional identity among engineers in the interwar period.

The discussion of papers was an important part of the IEI's activities from its inception. The Institution established an annual *Journal of the Institution of Engineers (India)* in its first year. The journal began by publishing a handful of papers every year, but by 1939 it had become a quarterly, and by the 1940s it featured papers, reports from the Institution's regional sections (discussed later), book reviews, and engineering news from around the world.[77] The IEI rapidly took on some of the roles the Industrial Commission had envisaged for it, advising the government and representing the Indian profession in international organizations for technical standards. These organizations worked for governments and industry, developing standards in order 'to secure interchangeability of parts, to cheapen manufacture ... and also to expedite delivery'.[78] In 1922 the IEI's Council became the Indian Committee of the British Engineering Standards Association (BESA; later British Standards Institution); the following year it was made the Indian National Committee for the International Electrotechnical Commission (IEC). The IEI continued to represent

---

all-branches principle appears to have become a definite part of the IEI's identity: the Institution continues to cover all branches of engineering to the present day.

[76] As the Charter put it, the IEI would '[facilitate] the scientific and *economic* development of Engineering in India' and 'encourage inventions and investigate and make known their nature and merits'. Annexure IV: 'Royal charter', in *IEI Marches On*, edited by Nandy. Emphasis mine.

[77] See the IEI *Journal* in the 1920s, for example, vol. 7 (December 1927), vol. 8 (April 1929), and vol. 9 (May 1930); IEI *Journal*, vol. 26, no. 2, part 1 (December 1945); Nandy, *IEI Marches On*, p. 27.

[78] C. LeMaistre, 'Summary of the Work of the British Engineering Standards Association', *Annals of the American Academy of Political and Social Science*, vol. 82, 'Industries in Readjustment' (March 1919): 247–52, here p. 247. LeMaistre was then Secretary of the BESA.

India on these bodies until after Independence, when the new Indian Standards Institution took over both functions.[79]

The IEI had the standard system of graded membership, the body corporate consisting of Members and Associate Members, who attended General Meetings and voted on decisions of the Institution.[80] While it has not been possible to identify the exact procedure by which members were admitted in the early years, it is almost certain that specific educational qualifications and experience were required, with the Council then voting on each case. (Annual reports, starting with the first year, refer to the number of members 'elected'.)[81] From 1928 onwards, candidates' theoretical knowledge was tested by the IEI's Associate Membership examinations.[82]

In fact, the Associate Membership examination was more than a way to select entrants to the Institution. It was also designed as a contribution to engineering education by creating a qualification for individuals who, though not in possession of a degree or diploma, were engaged in engineering work. This would have constituted an important opportunity for Indians in subordinate positions in industry who had not had the chance or the means to attend an engineering college. By 1939, passing Sections A and B of the Associate Membership examination had been accepted as equivalent to a degree by the governments of Punjab, Bengal, the United Provinces, the Central Provinces, Burma and Travancore, and by the Federal Public Service

[79] Institution of Engineers (India), Yearbook 1964–65 (Calcutta: Institution of Engineers, 1965), p. 6. Available at British Library (Shelfmark: General Reference Collection P.621/409).

[80] Annual Reports in IEI *Journal*, various years; Annexure IV: 'Royal Charter', in *IEI Marches On*, edited by Nandy, here pp. 85–6.

[81] See IEI *Journal*, vol. 2 (April 1922): 139; IEI *Journal*, vol. 18 (August 1938): 6–16.

[82] The original constitution or rules of the Institution for our period have not been found, but the 2012 bye-laws, which are the result of several successive amendments to those drafted in the 1930s, stipulate certain experience requirements and a possible interview in addition to the Associate Membership examination or equivalent qualifications. For Associate Members, the minimum age is twenty-six, and the candidate should have had 'at least five years professional engineering experience in a position of responsibility'. IEI Bye-Laws effective 5 May 2012, http://www.ieindia.info/PDF_Images/Bylaws/ByeLaws.pdf (downloaded 2 July 2012). See also 'The Royal Charter, the early years and the bye laws', Nandy, *IEI Marches On*, pp. 11–20.

Commission.[83] Thus, an engineer with IEI qualifications could apply for certain government engineering jobs even if he did not possess formal qualifications.[84] The IEI, in turn, exempted the holders of engineering degrees from certain universities from taking its A and B exams when they sought admission to the Institution.[85] Judging by the success rate of candidates, the Associate Membership exams were of a high standard. In the exams held in October 1933, only seven out of nineteen candidates passed Part A, while six out of seventeen passed Part B. The results in later years were similar. In 1936–7, fourteen out of twenty-nine candidates passed Section A of the exam; four out of fifteen passed Section B. In 1944–5 seventeen out of fifty-three passed Section A and nine out of thirty-six passed Section B.[86]

Under the provisions of its 1935 Charter (which had been 'vigorously pursued by Sir Rajendra Nath Mookerjee [President, 1920–1] and followed up strongly by succeeding Presidents'),[87] Members and Associate Members of the IEI were allowed to style themselves 'Chartered Engineer (India)'. They could also use the post-nominals MIE (Ind.) and AMIE (Ind.) respectively.[88] This form of recognition, or 'informal licensing' in Buchanan's words, was particularly important in a profession which did not require compulsory registration. As *Indian Engineering*—the Calcutta-based journal described earlier—commented in 1925 in the context of the British profession, anyone could 'put C. E. after his name, and call himself a civil engineer, if he is a plumber or a glazier or house-decorator or nothing at all.... But a *chartered* civil engineer is another affair, the designation implies certain qualifications'.[89] The corollary of the power to certify

[83] IEI *Journal*, vol. 20 (January 1941): 29.

[84] This was not limited to British India. For example, around 1946, the princely state of Travancore recognized the AMIE (Ind.) qualification for the purpose of recruitment to its engineering services. IEI *Journal*, vol. 26, no. 3 (March 1946): 69.

[85] IEI *Journal*, vol. 26, no. 3 (March 1946): 69.

[86] IEI *Journal*, vol. 15 (July 1935): 6; IEI *Journal*, vol. 18 (August 1938): 13; IEI *Journal*, vol. 26, no. 3 (March 1946): 68.

[87] Nandy, *IEI Marches On*, p. 11. For the year of Mookerjee's Presidency, see Annexure III, Nandy, *IEI Marches On*, pp. 75–7.

[88] Annexure IV: 'Royal Charter', Nandy, *IEI Marches On*, here p. 86.

[89] Lead article, *Indian Engineering*, 17 June 1925, pp. 351–2, here p. 352.

an engineer's competence was the need to police the conduct of members.[90] In August 1944 the IEI brought into force its 'Professional Conduct Rules' (Code of Ethics from 1954).[91]

The designation 'Chartered Engineer (India)' was also significant in that it indicated a specifically Indian professional identity—Indianness being defined not by the race of an engineer but by his geographical field of operation. Although it was not a legal requirement, the prestige of such a qualification must have been attractive to a young Indian engineer with no other affiliations. Other engineers (British and Indian) who were already members of the Empire-wide profession via the London institutions also found it advantageous to join the IEI, the MIE (Ind.) or AMIE (Ind.) badge perhaps indicating an additional India-specific expertise. Some engineers played leading roles in a 'home' institution as well as in the IEI, as in the case of R. D. T. Alexander, a member of the India Advisory Committee of the ICE who later became President of the IEI.[92]

From the start the IEI established a decentralized structure, with local centres in several Indian provinces 'for closer and efficient interchange [sic] of information and views'.[93] The membership of the Institution was distributed among these centres, presumably on

[90] Annexure IV, 'Royal Charter', Nandy, *IEI Marches On*, here p. 81 (paragraph 2[j]).

[91] Nandy, *IEI Marches On*, p. 33. In this sense, the engineering institutions in Britain and India were different from those for the medical profession—where one body (for example, the General Medical Council in Britain) licensed/regulated practitioners, and another (for example, the British Medical Association) represented the interests of practitioners. In the case of engineering, the same institution could represent the profession's interests and regulate it.

[92] Annexure III, Nandy, *IEI Marches On*, pp. 75–7; R. D. T. Alexander to Secretary of the ICE, 26 April 1926, part of the ICE's official correspondence of 1926 considering the creation of a local centre in India (ICE Archives, London: 185/03). Examples of Indians who were members of 'home' institutions and of the IEI are Fakirjee E. Bharucha, MIMechE., MIE (Ind.) and C. V. Krishnasawami Chetty, AMIEE, MIE (Ind.) (as of the late 1930s), both of whom are mentioned later in this chapter.

[93] According to B. N. Chaudhuri, MIE, 'A Short History of the Growth and Development of the Institution of Engineers (India)' in *Demicenturion* (IEI Commemorative volume [1969]), p. 43ff. The volume is available at the Library of the IEI, at its Kolkata Headquarters at 8 Gokhale Road. The IIC had envisaged the IEI as having local associations as the IMechE did in Britain (IIC Appendix H, pp. 386–7).

the basis of members' place of residence.[94] The local centres organized their own lectures, paper discussions, and visits to engineering works, in addition to the activities of the Institution as a whole.[95] As of 1937, there were local centres for Bengal, Bombay, southern India, the United Provinces, Mysore, north-western India, and even for members outside India.[96] The all-India membership increased steadily over the interwar years and beyond. The number of Corporate Members (Members and Associate Members) went from 138 in 1920 to 1,190 in 1940, before leaping to 4,168 in 1950; while the non-corporate membership (Honorary Members, Students, and so on) went from 192 in 1930 to 228 in 1940, and then soared to 3,140 in 1950.[97]

***

It must be noted here that the IEI's growth was not entirely unopposed. At the beginning, the Institution's eclectic character pitted it against one particular group of engineers in India: the PWD Engineering Congresses. J. W. Meares, a member of the IEI's first Council,[98] recalled in his 1934 autobiography that the Congresses had opposed the new institution fiercely:

> We realized very early that if this body was to take its place alongside the great engineering Institutions of England and America it must work entirely through 'local associations' in the various provinces. A nucleus existed in certain annual congresses held in [some] of the large centres, and we thought they would naturally welcome a co-ordinating body which would weld them all into a harmonious whole. The Government of India encouraged us, the Viceroy came to our inaugural meeting; and every one worked hard to ensure the success that will surely come presently, and is in fact now in sight. Nevertheless these isolated congresses one and all preferred at first to continue to plough their lonely and unproductive furrows,

[94] Membership totals in Annual Reports, IEI *Journal*, various years.

[95] See IEI *Journal*, vol. 18 (August 1938): 10.

[96] See Annual Report for year ending 31 August 1937, in IEI *Journal*, vol. 18 (August 1938): 6–16.

[97] Table in Nandy, *IEI Marches On*, p. 34.

[98] IEI *Journal*, vol. 1 (September 1921): 15.

unknown beyond the borders of their own cities, rather than join up and become powerful in the councils of India.[99]

In fact, the Congresses were hostile not because they wanted to remain confined to their provinces, but because they had their *own* vision of an all-India institution. As we have seen, the Punjab Engineering Congress had been hoping for some years to spearhead a pan-Indian PWD engineering movement of civil-engineer officers by federating the various provincial Congresses (and had made some progress in this direction). Understandably, it viewed the Calcutta-based IEI, headed by men such as geologist Thomas Holland, electrical engineer J. W. Meares, and engineer-businessman R. N. Mookerjee (a member of the IIC, 1916–18, and President of the IEI, 1920–1) as a usurper.[100] According to Meares, W. S. Dorman, the long-time Secretary of the Punjab Engineering Congress, 'wanted to continue bossing the show there', and the Congress 'did its utmost to queer the pitch for us [the IEI]'.[101]

Some members of the PWD Congresses also found the relatively inclusive IEI antithetical to their own elitist ambitions. J. W. Mackison, a municipal engineer in Bombay, recalled in 1926 that '[w]hen the recently formed Indian Engineering Society was started the Council of the Bombay Engineering Congress unanimously agreed, not to have anything to do with it'. Mackison claimed that the IEI had admitted several engineers whom the Congress had turned away. 'Many here,' he reported, 'regard [the IEI] as a back door for admission to the Institution of Civil Engineers.'[102] Strictly speaking, this could not

[99] John Willoughby Meares, 'At the Heels of the Mighty: being the Autobiography of "Your obedient humble servant"' (typescript, 1934), IET Archives London: SC 169/1/1, p. 236. Meares was, in the course of his career, Electrical Adviser to the Government of India and Chief Engineer, Hydroelectric Survey of India.

[100] For Mookerjee's membership of the Industrial Commission, see IIC *Report*, p. xvi; on his life and career, see K. C. Mahindra, *Rajendra Nath Mookerjee: A Personal Study* (Calcutta: Art Press [1933]).

[101] Meares, 'At the Heels of the Mighty', p. 397. Dorman, the founding Secretary of the Punjab Engineering Congress, was still Secretary as of 1920. See Montgomery, Presidential Address, pp. i–ii.

[102] J. W. Mackison to A. A. Biggs, 30 March 1926, supporting proposal for a Local Association of the ICE in India, ICE Archives, London: 185/03.

have been the case, as admission to the IEI did not automatically entitle one to membership of the London institutions. Mackison probably meant that whereas the PWD Congresses had wanted to gain the imprimatur of the ICE, the IEI now had that privilege. The 'back door' may also have been a reference to the fact that while the provincial Congresses were for gazetted (usually civil engineer) officers in the PWD and state railways, the IEI accepted engineers working in private industry, outside the gentlemanly paradigm of the PWD (which I discuss in Chapter 3).

However, as *Indian Engineering* pointed out, the PWD Congresses and the IEI were fundamentally different types of association—a congress was merely an annual meeting for the discussion of papers on engineering, and could not carry out the governing functions that the IEI did.[103] Whether out of recognition of this fact or the increasing evidence that the IEI had the blessing of the state, the PWD Congresses gradually reconciled themselves to cooperating with the Institution. As early as 1920, the President of the Punjab Congress, A. S. Montgomery, referred to the provincial Congresses' aim to combine into a nationwide organization, declaring that '[t]he seed thus sown has sprung up into the Institution of Engineers (India), and we may rightly claim, in my opinion, the majority of the credit for its creation'.[104] A few years later, G. H. Thiselton Dyer, President of the Bombay Engineering Congress, made a similar argument. He declared that his organization was not in competition with the IEI, going so far as to say that if the Congress had eventually to be dissolved due to dwindling membership, they would at least draw satisfaction from having been an early contributor to the establishment of an Indian engineering institution.[105] In some cases, important figures from the Engineering Congresses were also co-opted by the IEI. B. P. Varma, an irrigation engineer who had been influential in forming the Engineers' Association, Punjab (a forerunner of the Punjab

[103] Lead article, *Indian Engineering*, 17 June 1925, p. 352.
[104] Montgomery, Presidential Address, pp. i–ii.
[105] 'Engineering Congresses', *Indian Engineering*, 28 March 1925, pp. 169–70. The article refers to Thiselton Dyer's Presidential Address at the ninth Bombay Engineering Congress.
[106] See 'Presidential Address by Rai Bahadur B. P. Varma: President 1934–5', IEI *Journal*, vol. 15 (July 1935): 11–18.

PWD Congress), became President of the IEI in 1934–5;[106] while Sir Ganga Ram, a distinguished Roorkee alumnus, served on the Council of the IEI in 1922 and as President of the Punjab Congress the following year.[107] In 1943 the Bombay Engineering Congress was merged with the IEI.[108] The Punjab Congress, while cooperating with the IEI, remained a separate entity.[109]

Indianization and Industrialization

The IEI's distinguishing characteristic in the interwar period was its emphasis on Indianization and industrialization. This was referred to by Sir R. N. Mookerjee (1854–1936), President of the Institution for 1920–1, in his inaugural address. Mookerjee saw the formation of the IEI as heralding an era of equality, lauding 'the spirit of comradeship and cooperation in which British engineers have extended the hand of fellowship to their Indian colleagues'. He also saw the Institution as a means 'to promote the efficiency and training of Indian engineers', the better for them to harness India's plentiful natural resources in the cause of industrialization.[110]

In fact, Mookerjee's election as President in the first full year of the Institution's functioning was itself symbolic of the twin priorities of industrialization and Indianization. Unlike most of those who had represented Indian engineering in the past, he was not a colonial government officer. Instead, at a time when 'native' Indian engineers had limited opportunities, he had developed a successful career as

[107] 'History: Pakistan Engineering Congress'; IEI *Journal*, vol. 2 (April 1922): 137. On Ganga Ram's career, see 'Biographical Notes', appendix to K. V. Mital, *History of the Thomason College of Engineering (1847–1949): On Which Is Founded the University of Roorkee* (Roorkee: University of Roorkee, 1986), pp. 260–6, here p. 262.

[108] Nandy, *IEI Marches On*, p. 18.

[109] According to the website of the Punjab Congress's successor institution: 'For nearly thirty years with the Institution of Engineers India and later on with the Institution of Engineers Pakistan the stand taken by the Congress has been that it would welcome an association which does not amount to merger or loss of its own identity.' See 'History: Pakistan Engineering Congress'.

[110] Annexure II: 'The Inaugural Address', Nandy, *IEI Marches On*, pp. 65–74, here pp. 66, 70. The dates of Mookerjee's lifespan are mentioned in P. C. Mahalanobis, 'Sir Rajendra Nath Mookerjee: First President of the Indian Statistical Institute 1931–1936', *Sankhyā: The Indian Journal of Statistics*, vol. 2, part 3 (1936): 237–40.

an engineer-businessman. Starting as a contractor, he co-founded
the engineering firm Martin & Co. (later Martin Burn & Co.) in Calcutta.[111] He also took a strong interest in developing the education
and career prospects of his compatriot engineers. In 1910 he had suggested that a central technical college be created in India, rather than
sending Indians abroad for technical training. He was also instrumental in the opening of the East Indian Railway's Kanchrapara
workshop to Indian apprentices.[112]

While it has not been possible, in the absence of complete lists of
members, to establish the Indian share of the IEI's overall membership, the list of Mookerjee's successors as President indicates a gradual Indianization of the professional elite. Table 2.4 shows that of the
twenty-two Presidents until and including 1939–40, eight were Indian.
(In the following decade only one President was a European.)[113]

Indianization notwithstanding, the IEI was no revolutionary body.
As Table 2.4 shows, many Presidents were senior officials in the government services, mainly the PWD and Railways. Hugh W. Brady,
the first Secretary of the Institution, felt that 'a large proportion of
our membership will always be European'.[114] Certainly there continued to be several enthusiastic European members. As late as 1938,
the IEI reported that it had created a London office and a 'London
Committee' in order to enable 'members of the Institution on home
leave to get into touch with each other and with the Institution, and
also to arrange for the Annual Institution Luncheon'.[115] Nevertheless,
Presidents of the IEI in Calcutta often placed emphasis on the Indian
members in their annual addresses, initially in somewhat paternalistic terms. W. H. Neilson (President, 1926–7), speaking in the context
of the low number of papers submitted to the Institution in its early

[111] 'Life and Work of Sir Rajendra Nath Mookerjee: The Inaugural-President of The
Institution of Engineers (India)' in *The Institution of Engineers (India), Diamond Jubilee
1980: Souvenir* (Calcutta, 1980), pp. 35–7; K. C. Mahindra, *Rajendra Nath Mookerjee: A
Personal Study* (Calcutta: Art Press, [1933]).

[112] 'Life and Work of Sir Rajendra Nath Mookerjee'.

[113] The assumption here is that none of the European names refers to an Anglo-
Indian.

[114] Hugh Brady, 'The Origin, Character and Prospects of the Institution', IEI
Journal, vol. 1 (September 1921): 30–7, here p. 37.

[115] IEI *Journal*, vol. 18 (August 1938): 10.

Table 2.4 Presidents of the Institution of Engineers (India) (IEI), 1920–40[*]

Year	Name	Field of work (if known)	Designation at the time of presidentship (if known)	Source for field/ designation
1920 (before formal registration)	Sir Thomas R. J. Ward	PWD	Inspector–General of Irrigation, India. (Retd. Jun 1921)	*India Office List (IOL)* for 1940
1920–1	Sir Rajendra Nath Mookerjee	Private industry	Co-founder, Martin & Co., Calcutta	Biography by K. C. Mahindra[†]
1921–2	Col. Sir George Willis	Military	Mint Master, Bombay	*IOL* 1940
1922–3	A. C. Coubrough			
1923–4	Sir Clement D. M. Hindley	Railways	Chief Commissioner, Railway Board	*IOL* 1940
1924–5	H. Burkinshaw			
1925–6	Dewan Bahadur A. V. Ramalinga Aiyar	PWD		*IOL* 1920
1926–7	W. H. Neilson			
1927–8	Sir James S. Pitkeathly	Indian Stores Dept.	Chief Controller of Stores	*IOL* 1940
1928–9	Lt. Col. R. D. T. Alexander			
1929–30	Lt. Gen. Sir Edwin H. de Vere Atkinson	Military	Master-General of Ordnance	*IOL* 1940
1930–1	C. Addams Williams	PWD	Chief Engineer	*IOL* 1940
1931	Diwan Bahadur N. N. Ayyangar	PWD		*IOL* 1940

(Cont'd)

Year	Name	Field of work (if known)	Designation at the time of presidentship (if known)	Source for field/ designation
1931–2	Raja Jwala Prasad	PWD	Chief Engineer; retd. Nov 1931	*IOL* 1940
1932–3	Dr A. Jardine	Private industry	a director of Jessop & Co.,engg. contractors	Obituary in ICE *Proceedings*‡
1933–4	Thomas Guthrie Russell	Railways	Chief Commissioner, Railway Board	IOL 1940
1934–5	B. P. Varma	PWD		IEI *Journal*§
1935–6	F. C. Temple			
1936–7	Rai Bahadur Chhuttan Lal	PWD	Chief Engineer; retd. 1937)	*IOL* 1940
1937–8	Fakirjee E. Bharucha	Pvt. indus-try (textile mills)		IEI *Journal***
1938–9	E. J. B. Greenwood	PWD	Govt. Electrical Inspector, Madras, as of 1920	*IOL* 1920
1939–40	Khan Bahadur M. Abdul Aziz	PWD	Chief Engineer rank	*IOL* 1940

Notes: * Dates and names of Presidents are given in Annexure III, H. Nandy (ed.), *IEI Marches On* (Kolkata: Cdr. A.K. Poothia for Institution of Engineers (India), 2002 [1996]), pp. 75–7. Their occupations were ascertained from various sources, listed in the table itself.

† K. C. Mahindra, *Rajendra Nath Mookerjee: A Personal Study* (Calcutta: Art Press [1933]).

‡ 'Alexander Jardine, DSc', obituary appearing in ICE *Proceedings*, vol. 48, no. 4 (April 1971): 723–4.

§ See 'Presidential Address by Rai Bahadur B. P. Varma: President 1934–5', IEI *Journal*, vol. 15 (July 1935): 11–18.

** 'Presidential Address by Mr. Fakirjee E. Bharucha, M.I.Mech.E., M.I.E. (Ind.): President 1937–38', IEI *Journal*, vol. 18 (August 1938): 17–30.

years,[116] said: 'I would impress upon members, more particularly our Indian members, for whom the Institution was primarily founded, the necessity, and advantage, of submitting papers, on subjects with which they are most conversant'.[117] E. H. de Vere Atkinson (President, 1929–30) acknowledged that 'this is the country of the Indian', although he qualified his support for Indianization by saying that it should proceed 'on the right lines': an engineer must not be selected to a post unless he was competent to occupy it.[118] Atkinson's words appear to reflect, albeit mildly, contemporary British reservations about the suitability of Indian engineers for responsible work (these attitudes and the reasons behind them are discussed in Chapters 3 and 4).

In contrast, a mildly nationalistic note was struck by an Indian President, Jwala Prasad (1931–2), a Roorkee alumnus who had become Chief Engineer in the United Provinces PWD some years before:[119]

An earlier predecessor and that an original founder stated on a similar occasion that this Institution was primarily meant for Indians. By this I think he did not mean an engineer in India but an Indian engineer. As I belong to this category it is but natural that my sentiments and opinions but follow the trend of the blood in my veins.[120]

Drawing attention to the Hindu epics and to recent archaeological discoveries, Prasad argued that ancient India had a rich history of engineering. He referred to

the construction of the famous bridge over the sea at Cape Comorin ... the flying of Rama to Ajodhya in a single day after the conquest of Lanka to save his devoted brother Bharat ... the cutting of the Gangotri from a wonderful glacier through disinfecting rocks and land by [Rama's] great

[116] In volumes 7 (1927), 8 (1929), and 9 (1930), only three papers each were published in the Institution's *Journal*.

[117] Address by W. H. Neilson, IEI President (1926–7), IEI *Journal*, vol. 7 (December 1927): 9–24, here p. 9.

[118] Presidential Address by Lieut-Genl Sir Edwin H. de Vere Atkinson (President, 1930), IEI *Journal*, vol. 10 (May 1931): 10–16, here p. 11.

[119] Record of Services, *India Office List* for 1938.

[120] Presidential Address by Jwala Prasad, IEI *Journal*, vol. 12 (July 1933): 19–23, here p. 19.

ancestor Bhagirath, before men knew how to dig a well. Some people dismiss these ideas with a sneer but be it remembered that our ancestors had a great reputation both for imaginative construction and veracity, that the Romans, whose remarkable works still remain, acknowledge their kinship with Indian culture and that the diggings at Sarnath, [Mohenjodaro] and other places cannot otherwise be accounted for.[121]

The other recurring theme in the IEI's discussions was Indian industrialization. This reflected one of the institution's other goals: to encourage industrial engineers. While traditional areas like civil/ public works engineering continued to receive attention, papers discussed at the Institution's meetings were not limited to these fields, but included a range of industrial topics too (see Table 2.5).

Several Presidents, including some who were senior government officers, deplored the traditional mindset of engineers who set their minds on government jobs instead of working in private industry. In his presidential address for 1929–30, Edwin Atkinson held up the former President R. N. Mookerjee as an exemplar of the industrial engineer. 'I fear,' he said, 'that even to-day every Engineering student looks

Table 2.5 Titles of some papers discussed at the IEI meetings

Author(s)	Title of paper	Details of publication in IEI *Journal*
A. R. Beattie	The B.T.U. in an Indian Paper Mill	Vol. 15 (July 1935)
Goverdhan	Cooling Water for Diesel Engines	Vol. 15 (July 1935)
J. W. Meares	The Possibility of Flood Regulation and Conservation in the Himalayas for Irrigation or Power	Vol. 15 (July 1935)
M. L. Garga	Design and Construction of Tinai Nadi Aqueduct (Sarda Canal)	Vol. 18 (August 1938)
M. S. Bhandarkar and Prof. K. Aston	Electrical Manufacturing Industry in India and the Scope and Line of its Future Growth	Vol. 20 (January 1941)
Dr M. A. Korni	Rontgenology in Reinforced Concrete	Vol. 20 (January 1941)

Source: Compiled by the author.

[121] Presidential Address by Jwala Prasad, p. 22.

forward only to Government employment.... It is by private initiative that the Engineering profession can best foster the industrial prosperity of India.'[122] Fakirjee E. Bharucha (MIMechE, MIE [Ind.], IEI President 1937–8), who, beginning in the Bombay textile industry, had spent his working life as a mechanical engineer, devoted his address to a discussion of the state of mechanical (and by extension industrial) engineering in the country. Fifty years ago, he said, mechanical engineering had been seen as an occupation for 'those young men who were dunces and backward in their scholastic career,... as they were considered unfit for the so-called noble professions of law, medicine and civil engineering'. Such 'aristocratic prejudice' notwithstanding, Bharucha argued, mechanical engineering, by virtue of its close connection with industry, had become 'the most important branch of engineering without which no civilized country could thrive and have a place of honour in this world'.[123] Mechanical engineering was no more restricted to 'the upkeep of prime movers of factories or management of mechanic shops'; there was now a need for 'scientifically trained engineers' who could also design, manufacture, and maintain the machinery in use in factories.[124] Unfortunately, the education system was not geared towards producing industrial engineers.[125]

Some Indian office-bearers pursued the theme of industrial development in more emphatic, sometimes nationalistic, terms. In the late 1930s, M. C. Bijawat, an irrigation engineer and Chairman of the United Provinces Centre of the IEI, drew a connection between personal courage and the pursuit of industrial engineering. Government jobs in engineering were scarce, he said, and 'it is a pitiable sight indeed to find a handful of products of a few Engineering Colleges begging from door to door for employment'. Engineers

must no longer hanker after Government services, a hankering which exercises a very cramping effect on the intellectual development of the students and roots out their self-reliance. They should take up the line in a truly professional spirit which will give full scope for the exercise of their creative

[122] 'Presidential Address by Lt-Genl Sir Edwin H. de Vere Atkinson', p. 11.
[123] 'Presidential Address by Mr Fakirjee E. Bharucha, MIMechE., MIE (Ind.): President 1937–38', IEI *Journal*, vol. 18 (August 1938): 17–30, here p. 17.
[124] 'Presidential Address by Mr Fakirjee E. Bharucha', p. 24.
[125] 'Presidential Address by Mr Fakirjee E. Bharucha', pp. 27–8.

genius and to develop their initiative, they must learn to put the knowledge they have gained in their colleges to practical use in furthering research in all branches of the profession and help in making new discoveries and inventions.... They will thus not only find employment for themselves, but will be able to create it for millions of other people... [126]

Although the politics of IEI members varied, the discourse on industrialization became closely linked with economic nationalism, particularly from the late 1930s. The historian Gyan Prakash, writing on the link between the colonial state's use of technology and the growth of nationalism, has argued that '[w]hat began as an effort to relocate colonial power in technical apparatuses and practices unleashed a political struggle to establish a nation-state that would institute the logic of rational artifice more fully and efficiently'.[127] In a similar way, Indian engineers, while still operating within the colonial economy and state institutions, felt that through industrialization they could play an important role in the making of a self-reliant India. C. V. Krishnaswami Chetty (AMIEE, MIE [Ind.]), Chairman of the South India Centre of the IEI, said in 1939 that since the outbreak of World War II, Indians had been made aware of

> their utter dependence on other countries for most of their daily wants. It has opened the eyes of the people to the urgent necessity for starting various industries and especially key-industries [sic]. If the impetus now given to industrialization is not taken advantage of, I am afraid, India will ever be dependent on other countries for most of her wants [except] food.[128]

Referring to the National Planning Committee (set up under the aegis of the Indian National Congress, which then led a large number of provincial governments, in anticipation of political autonomy at the national level), Krishnaswami Chetty declared that the IEI must offer its support. 'Our Institution must place at the disposal of the National

---

[126] Address of M. C. Bijawat, Chairman of the United Provinces Centre of the IEI, in IEI *Journal*, vol. 18 (August 1938): 46–54, here pp. 53–4.

[127] Gyan Prakash, *Another Reason: Science and the Imagination of Modern India* (Princeton: Princeton University Press, 1999), p. 199.

[128] Address by Rao Bahadur C. V. Krishnaswami Chetty (Chairman, South India Centre) on 2 December 1939, in IEI *Journal*, vol. 20 (January 1941): 87–91, here p. 89.

Planning Committee all the techincal [*sic*] assistance it can give. Our Institution must play an important part in the industrialization of the country.... Time has come for engineers to evince greater interest and take part in the political life of the country.'[129]

The cause of Swadeshi (indigenous) industry and materials was also taken up. Fakirjee Bharucha insisted in 1938 that 'there is no sense in sending our raw materials to be manufactured into finished articles for us in foreign lands, when we could do it ourselves'.[130] In the Annual General Meeting of 1945, S. B. Joshi proposed a reso-lution advocating the use and manufacture of 'SWADESHI mate-rial, plant and equipment, without regard to cost and quality, with a view to advancing the Science of Engineering in India'.[131] When other members suggested that the clause 'without regard to cost and quality' be amended, Joshi argued that this would render the resolu-tion meaningless; he was specifically suggesting that engineers buy Indian material even when it was *not* the most economical alterna-tive. After all, it went without saying that if the quality was sufficient, engineers should buy Indian. Anyone who did not 'is not an Indian, he is an enemy'.[132]

***

This chapter has identified several key transformations occurring in the engineering profession in India over the period 1900–47. Over this period, the model of a profession dominated by government-officer engineers in the public works, railways, and military was altered con-siderably, as engineering activity expanded and diversified to include mechanical and electrical engineers working in large-scale industry. Equally importantly, the racial composition of its practitioners under-went a substantial shift. The profession, dominated by expatriate British engineers in the nineteenth century, included an increasing proportion of 'native' Indians in the twentieth. These transforma-tions should be seen as a part of the evolution of the economic and

---

129 Address by Rao Bahadur C. V. Krishnaswami Chetty, p. 90.

130 'Presidential Address by Mr Fakirjee E. Bharucha', p. 29.

131 IEI *Journal*, vol. 26, no. 2, part 2 (December 1945): 32.

132 IEI *Journal*, vol. 26, no. 2, part 2 (December 1945): 33.

political relationship between Britain and India after World War I, when industrialization and Indianization became important items of government policy. The development of professional institutions reflected these changes, and offers a fruitful lens through which to understand them.

An equally important, and related, transformation occurred in the identities that engineers in India fashioned for themselves. Membership of the London institutions, especially in the long nineteenth century, was a mark of belonging to a wider imperial, or international, engineering community. But the formation of the Institution of Engineers (India) in Calcutta in 1920 represented the beginnings of the idea of an Indian profession independent of the metropolis. Like the London Institutions did for the British profession, the IEI monitored the qualifications, performance, and conduct of its members, while providing a forum for them to exchange ideas on technical topics and matters relating to the profession. Both Europeans and Indians working in India joined the IEI, and some of them were also members of the London Institutions; but the IEI's focus was on matters pertaining to India. In particular, it encouraged industrialization and Indianization in the profession. It offered qualifications equivalent to degrees for practitioners who had not been formally trained, afforded its Corporate Members recognition through the title of Chartered Engineer (India), and represented India in imperial and international forums. Although the Institution, as the brainchild of a government-appointed Commission, represented a top-down approach to creating an Indian profession, the Indian identity acquired a momentum of its own within the Institution's framework. This was reflected in the rhetoric of the IEI's office-bearers, both European and Indian, which included increasingly confident expressions of Indian ability, and in which a form of economic nationalism became visible from the late 1930s.

However, it should not be concluded from this story that the new Indian identity emerged uncontested. It was opposed vigorously in the beginning by the PWD Engineering Congresses operating in various provinces, which held a more exclusive vision for the engineering profession, centred on civil engineering, government-officer status, and the (hoped-for) approval of the Institution of Civil Engineers in London. Again, it would be simplistic and inaccurate to view this as a conflict between the votaries of colonialism and nationalism. In fact, the emergence of the IEI as the more prominent representative of Indian engineering was due not only to its inclusiveness but also to the endorsement it received from the colonial state—a state whose nature and priorities were undergoing important changes in the inter-war period. Yet recognition did not come easy for Indian engineers, whose suitability for responsible positions continued to be a matter of debate within the organizations that employed them. The differential professional experiences of British and Indian engineers in various sectors are analysed in detail in the chapters that follow.

CHAPTER THREE

Men of Character

Indianization and the Culture of Public Works Engineering

It is entirely due to British engineers and British capital that we have built up what is one of the greatest boons and blessings to the people of India and a bulwark against the recurrent famines from which India has suffered in the past. We have no right to take the risk that such a state of affairs might be brought about again, either partially or to a very great extent due to the lowering of the efficiency of this irrigation service. We feel most strongly that it is absolutely essential that the English officials should be continued in the irrigation service.

—Vice-Admiral Taylor, MP, in the House of Commons (1935)[1]

Established in 1854, the Indian Public Works Department (PWD) was one of the most important departments of the colonial government, engaged in the construction and maintenance of roads, bridges, buildings, and irrigation works.[2] Its engineer officers, organized into

---

[1] 'Clause 233. (Services recruited by Secretary of State)', House of Commons Debate, 4 April 1935, vol. 300, cc. 582–635, here c. 616 (Hansard).

[2] For a nineteenth-century description of the functions of the PWD, see C. D. Maclean (ed.), *Manual of the Administration of the Madras Presidency: In Illustration of the Records of Government and the Yearly Administration Reports*, reprint, 3 vols (New Delhi: Asian Educational Services, 1987 [1885–93]), vol. 1, pp. 365ff. As mentioned in Chapter 2, the state-run railways also came under the PWD until 1905. The present-day Central Public Works Department of India gives 1854 as the date when the central government first had a 'Secretary of the Department of Public works'. See paragraph 13 of 'History of C.P.W.D.', document available under the tab 'Organisation', 'Historical Background' at http://cpwd.gov.in (accessed 21 July 2012).

large all-India and provincial bureaucracies (or 'services'), were at the executive and administrative heart of the Department, and were employed in substantial numbers (the all-India engineers alone numbered around 675 in 1929).[3] In the years 1900–40, growing demands for Indianization and the introduction of the principle of provincial autonomy necessitated a rethinking of how the PWD's engineering services should be organized. This chapter demonstrates that while policymakers in Britain and in India recognized that reorganization was required, they sought to upset as little as possible the existing culture of the engineering services. I argue that their debates and decisions revealed prevailing ideas of the qualities required of the ideal public works engineer.

Despite their importance, public works engineers in twentieth-century India, and their bureaucratic organization within the PWD, have received little attention from historians. Concentrating on the role of public works—dams, railways, and the telegraph—in the entrenchment of the colonial state, the literature has been concerned primarily with the second half of the nineteenth century, when Crown replaced Company. The rise of public works in this period is seen as 'an important step in situating the exercise of colonial power in the exploitation of the territory and the people as resources'.[4] These works also provided a legitimation of the colonial state. 'In stone, steel and steam they embodied the idea of the British Raj as a technological empire, able by its grand works and feats of engineering to master forces of nature that had defied and enslaved Indians for centuries.'[5] Some scholars have studied specific aspects of the PWD such as its close relationship with the Indian engineering colleges and with the

---

[3] Estimate based on lists of officers of the Indian Service of Engineers (ISE) in the *India Office List* for 1929. The source lists separately the officers posted in each province/presidency. In addition, there was a lower service of engineers employed by the respective provinces, whose combined strength would have been several hundred more (based on K. V. Mital's history of the Roorkee Engineering College, cited later in this chapter).

[4] Gyan Prakash, *Another Reason: Science and the Imagination of Modern India* (Princeton: Princeton University Press, 1999), chapter 6. Quoted text on p. 162.

[5] David Arnold, *Science, Technology and Medicine in Colonial India*, The New Cambridge History of India, III.5 (Cambridge: Cambridge University Press, 2000), p. 121.

military.[6] Others have been more directly concerned with the PWD's engineers and their ideology: David Gilmartin argues that in the decades after the 1880s (when a network of irrigation canals was constructed in the Indus Basin), the engineering solutions ('imperial science') of irrigation engineers came into conflict with, and were checked by, civil administrators' 'science of empire', a theory of good governance that took into account notions of local community, custom, and hereditary rights.[7]

Overall, however, there is no detailed account of the internal structure of the PWD (which was essentially made up of engineers and engineering subordinates[8]) as a whole, or how it was affected by inter-war constitutional changes in India. In particular, little is known of the career patterns, organization, and professional identities of the engineers of the PWD who were engaged in the design and building of dams, canals, bridges, and roads.[9] This historiographical neglect of the PWD's engineering service contrasts with the study of other government services. The Indian Civil Service (ICS), its modes of recruitment, and the attitudes and worldviews of its officers have been a

[6] Arun Kumar, 'Colonial Requirements and Engineering Education: The Public Works Department, 1847–1947' in *Technology and the Raj: Western Technology and Technical Transfers to India 1700–1947*, edited by Roy MacLeod and Deepak Kumar (New Delhi, Thousand Oaks, and London: Sage Publications, 1995), pp. 216–32; John Black, 'The Military Influence on Engineering Education in Britain and India, 1848–1906', *The Indian Economic and Social History Review*, vol. 46, no. 2 (2009): 211–39.

[7] David Gilmartin, 'Scientific Empire and Imperial Science: Colonialism and Irrigation Technology in the Indus Basin', *The Journal of Asian Studies*, vol. 53, no. 4 (November 1994): 1127–49. Also on the ideology of irrigation engineers, but for a different period (the 1940s and 1950s), see Daniel Klingensmith, *'One Valley and a Thousand': Dams, Nationalism, and Development* (New Delhi: Oxford University Press, 2007), chapter 5.

[8] For example, see the PWD listing for Madras Presidency, *India Office List* for 1924, p. 35, and the more detailed listing for the PWD in that presidency in 1922: Govt of Madras, PWD, *Classified List and Distribution Return of Establishment, corrected up to 30th June 1922* (Madras: Superintendent, Government Press, 1922) [APAC: IOR/V/13/831]. (APAC refers to the Asia, Pacific and Africa Collection, British Library, London.)

[9] Some partial exceptions are Arun Kumar, 'Colonial Requirements'; Black, 'The Military Influence'; and Daniel R. Headrick, *The Tentacles of Progress: Technology Transfer in the Age of Imperialism, 1850–1940* (New York and Oxford: Oxford University Press, 1988), chapter 9. These works deal with engineering/technical education in relation to the PWD.

particularly fertile field of historical research,[10] while the Indian Medical Service (IMS) in the long nineteenth century and the scientific services of the government in the interwar years have also received historical attention.[11]

This chapter examines the work culture and organization of the Indian PWD's engineering services in the period 1900–40. In it I explore the following questions: What functions was the typical engineer-bureaucrat in the PWD expected to carry out? How did the training of engineers and systems of recruitment relate to the bureaucratic structure of the PWD, and how did these systems change over time in response to the progressive transfer of administrative power from appointed colonial officials to elected Indian ministers in the provinces?

The chapter is divided into two main sections. Starting in 1900, when the PWD was dominated by Europeans, and extending to the outbreak of World War II, when Indians made up more than 50 per cent of its elite engineering service, the first section describes the changing patterns of training, recruitment, and organization of PWD engineers. The second section explains some of these changes, and officials' response to them, in relation to the culture and functions of PWD engineering.

AN EVOLVING BUREAUCRACY: RECRUITMENT AND REORGANIZATION

The Imperial and Provincial Services, 1900–20

As of 1900, the PWD was a department of the Government of India with its own secretariat at the centre and a member of the Viceroy's

[10] On the ICS, see David C. Potter, *India's Political Administrators 1919–1983* (Oxford: Clarendon, 1986); C. J. Dewey, 'The Education of a Ruling Caste: The Indian Civil Service in the Era of Competitive Examination', *The English Historical Review*, vol. 88, no. 347 (April 1973): 262–85; Maria Misra, 'Colonial Officers and Gentlemen: The British Empire and the Globalization of "Tradition"', *Journal of Global History*, vol. 3, no. 2 (July 2008): 135–61.

[11] On the IMS, see Mark Harrison, *Public Health in British India: Anglo-Indian Preventive Medicine 1859–1914* (Cambridge, New York, and Melbourne: Cambridge University Press, 1994). For the scientific services, see Arnold, *Science, Technology and Medicine in Colonial India*, chapter 5.

Council at the helm. Under him were Secretaries and Under-Secretaries to government, consulting engineers or advisers to government, and Inspectors of engineering. The actual work of the PWD was done in the provinces, each of which had its own PWD Secretariat headed by one or more Chief Engineers. Most provincial PWDs were further divided into two branches: Roads and Buildings, and Irrigation.[12] The structure of the PWD in each province as of 1900, along with the system of recruiting engineers at different levels, is summarized in Figure 3.1.

As the figure shows, the engineer officers of the PWD were divided into two services: the Imperial Service of Engineers, whose members were recruited in Britain, and the Provincial Service of Engineers, with its members recruited in India. Members of a service were subject to a common set of rules regarding pay, leave, and promotion, irrespective of their physical location (the work of an engineer, whether Imperial or Provincial, lay in the province where he was posted). The main distinction between the Imperial and Provincial Services lay in these terms and conditions, not in the nature of work or the ranks that they could attain.[13] Engineers of the Imperial Service enjoyed higher pay, more favourable leave arrangements, and greater prestige than their Provincial counterparts (discussed in detail later).

Recruitment to the Imperial Service of Engineers was controlled by the Secretary of State for India, based at the India Office in Whitehall.[14] There were two main sources of manpower, each supplying a

[12] Based on Government of India, PWD, *Classified List of Establishment* for various years. Available in APAC, shelfmarks in the series IOR/V/13. See also *India Office List* for various years.

[13] *Royal Commission on the Public Services in India. Report of the Commissioners*, vol. I (Cd. 8382, London: HMSO, 1917), p. 19 (paragraph 25). This report is hereinafter cited as *Islington Commission Report*.

[14] The following paragraphs on the sources of recruitment are based on the following works (except when another source is specifically cited): Arun Kumar, 'Colonial Requirements', esp. p. 228; Brendan P. Cuddy, 'The Royal Indian Engineering College, Cooper's Hill, (1871–1906): A Case Study of State Involvement in Professional Civil Engineering Education' (PhD thesis, London University, 1980), esp. chapter 2 and pp. 147–8; K. V. Mital, *History of the Thomason College of Engineering (1847–1949): On Which Is Founded the University of Roorkee* (Roorkee: University of Roorkee, 1986), p. 78; Headrick, *Tentacles of Progress*, chapter 9; Annexure XVIII: 'Public Works Department and Railway Department (Engineering Establishment)', *Islington Commission Report*,

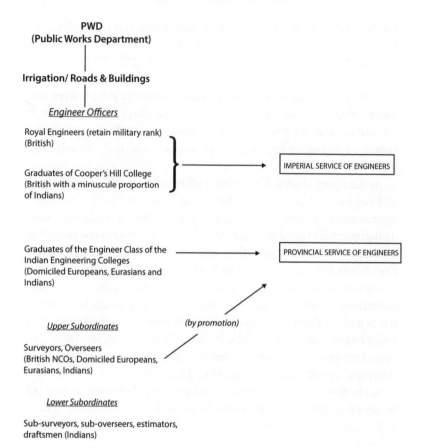

Figure 3.1   Public Works Department (PWD) organization and recruitment at 1900
Source: Compiled by the author from several sources.

fixed quota of engineers annually. The first of these was the military in
Britain: six Royal Engineers, trained in the military engineering colleges of Addiscombe, Woolwich, and Chatham, were selected annually for the PWD. These men, who retained their military rank and
privileges, represented the continuation of a tradition: public works
in the early years of the PWD, and indeed before it was founded, had

pp. 326–37, esp. p. 329; Hirday Nath Kunzru, *The Public Services in India (Political
Pamphlets—II)* (Allahabad: Servants of India Society, 1917), p. 146.

been carried out mainly by military officers. Their continuing employ-ment in civil positions was not confined to the PWD, but extended also to the state railways.[15]

The second source of engineers for the Imperial Service was the Royal Indian Engineering College (RIEC) at Cooper's Hill, from which fifteen graduates were selected annually. This college had been set up near London in 1871 for the express purpose of satisfying the Indian PWD's demand for engineers, and its expenses were borne by the Government of India. The RIEC admitted students on the basis of an entrance examination that tested them on English, mathematics, Latin, Greek, French, German, natural and experimental sciences, and mechanical and freehand drawing (English and the other languages made up close to 50 per cent of the total marks). The curriculum at the College itself covered engineering, mathematics, natural science, language (including Hindustani), and a period of apprenticeship to a practising engineer. Students were also instructed in the history and geography of India. This system of recruitment meant that the Impe-rial Service of Engineers was almost entirely made up of Britons, the only exception being a select few Indians who could afford to travel to England and study at Cooper's Hill. By a rough estimate, Indians con-stituted well under 5 per cent of the graduates of Cooper's Hill over the duration of the College's existence.[16] The Provincial Service, on the other hand, was composed of Europeans domiciled in India, Eura-sians, and Indians. Indeed, its creation in 1895 had been advertised as a means for granting greater opportunities for Indian engineers to join the PWD. Engineering officers were recruited from the Engineer Class of the Indian engineering colleges at Roorkee, Madras, Poona, and Sibpur near Calcutta. These had been established from the mid- to late nineteenth century, and also ran classes for Upper and Lower Subordinates, whose graduates joined the PWD at the corresponding subordinate levels. Nine and ten engineers in alternate years were appointed to the Provincial Service, the largest share coming from the oldest of the Indian colleges, Roorkee's Thomason College of

[15] See Chapter 4.

[16] This estimate is based on Appendix C: 'Students of the R.I.E.C.' in Cuddy, 'Royal Indian Engineering College', pp. 310–37. The uncertainty is primarily due to the difficulty of distinguishing some Anglo-Indian names from European ones.

Engineering. In addition to this, five and four Upper Subordinates were promoted to the Provincial Service of Engineers in alternate years (in a given year the total number of appointments to the Provincial Service was fourteen).

The PWD divided each province into geographical areas of responsibility called circles, divisions, and subdivisions. The lowest rank in either Service (Imperial or Provincial) was Assistant Engineer (AE), responsible for a subdivision. Executive Engineers (EE), the next higher rank, were placed in charge of a division. Above this were the two 'administrative' ranks, their main functions being supervisory. These were Superintending Engineer (SE, in charge of a circle) and Chief Engineer (CE). There were usually one or two Chief Engineers in a province, one in charge of the Roads and Buildings branch for the whole province, and the other heading Irrigation.[17]

This system of recruiting for the PWD by patronage given to specific colleges and the Royal Engineer Corps was quite unlike the competitive examination system used for the Indian Civil Service.[18] It was also under strain. The Cooper's Hill College was in a precarious position at the turn of the twentieth century. In spite of the high fees it charged its students, it was not financially self-sustaining—a fact resented by the Government of India, which bore its expenses and against whose initial wishes the College had been set up. The College also faced opposition from other British colleges and universities that had developed degree-level courses in engineering in the latter half of the nineteenth century. These institutions argued that their students should be eligible for appointment to the PWD. Matters came to a head when the RIEC's last President, in a desperate attempt to reduce costs, summarily dismissed several long-serving faculty members. The Board of Visitors conducted an inquiry, and in 1904 the Secretary of State's Council of India voted for the abolition of the college. It was finally closed in 1906.[19]

[17] Annexure XVIII: 'Public Works and Railway (Engineering) Departments', *Islington Commission Report*, p. 326 (paragraph 2) (see pages 332–3 for the usage of the term 'administrative grades'); PWD *Classified Lists*.

[18] For more on the systems of competitive examinations and nomination/patronage, see Chapter 1.

[19] Cuddy, 'Royal Indian Engineering College', pp. 265–92.

The abolition of the RIEC did not mean an increase in recruitment from the Indian engineering colleges. Instead, the India Office continued to recruit civil engineers in England, turning now to the colleges that had earlier opposed the RIEC's monopoly. Men between the ages of twenty-one and twenty-four who had either obtained one of a specified list of British engineering degrees or passed the Associate Membership examination of the Institution of Civil Engineers were eligible to apply for recruitment in London. These qualifications included the BSc (Engineering) courses in the Universities of London, Glasgow, and Edinburgh; Sheffield's BE; Liverpool's BEng; and Cambridge's BA Honours (Mechanical Sciences Tripos). Candidates would be interviewed at the India Office and their suitability in terms of health, riding ability, and 'moral character' assessed. Those selected would be appointed to the post of Assistant Engineer in the Imperial Service of Engineers and posted to one of the Indian provinces.[20] While the India Office did not initially insist on candidates having had prior experience on engineering projects, they did consider it important, and by the late 1910s the regulations stated explicitly that candidates should preferably have undergone 'at least one full year's practical experience of Civil Engineering under a qualified civil engineer' in addition to their college training (or three years' practical experience if they had not undergone formal instruction in a college).[21] As of 1912, the Selection Committee was made up of one member of the Secretary of State's Council of India and two engineers, one of whom was a nominee of the Institution of Civil Engineers.[22]

Thus, recruitment in Britain for the Imperial Service continued despite the closure of Cooper's Hill. In fact, the number of London

[20] See 'Indian Public Works Department. Regulations as to Appointments of Assistant Engineers, 1910' and Appendix I in the *India Office List* for 1910, pp. 222–3.

[21] 'Indian Public Works Department and Indian State Railways. Regulations as to Appointment of Assistant Engineers in 1919' in file titled 'Regulations Governing Appointments of Engineers, with Selection Committees' Reports' [APAC: IOR/L/PWD/5/29]. The 1907 regulations only state that 'it is desirable that candidates should have had some experience as assistant in the preparation of the designs for, or in the execution of, some engineering work of importance'. 'Indian Public Works Department. Regulations as to Appointment of Assistant Engineers, 1907', in 'Regulations Governing Appointments of Engineers' [APAC: IOR/L/PWD/5/29].

[22] Annexure XVIII, *Islington Commission Report*, p. 330.

recruits now went up, the annual average being thirty between 1909 and 1914.[23] As in the Cooper's Hill days, Indians could apply for Imperial Service positions, but they would 'be selected to the [maximum] extent of 10 per cent' of the India Office recruits in a given year, 'if duly qualified'. Outside of this 10 per cent limit, '[e]very Candidate ... must be a British subject of European descent, and at the time of his birth his father must have been a British subject, either natural-born or naturalised in the United Kingdom'.[24] As is apparent from this, only a very small number of Indians were recruited in London. Three Indians were recruited by the Secretary of State in 1909,[25] and even then the Selection Committee was constrained to 'record their opinion that all the Indian candidates would, had they been Europeans, have been unhesitatingly rejected. They have the necessary academic qualifications, it is true, but with perhaps the exception of [one of the successful candidates], they have no experience of any practical value'. In most subsequent years the India Office continued to select three Indians,[26] although the Committee was remarking by 1912 that there was 'a marked improvement' in 'the standard attained by the native candidates who appeared before us'.[27]

The sanctioned strengths of the Provincial and Imperial Service were approximately in the ratio of 3:7.[28] Yet it appears that the proportion of Provincial engineers was in practice even lower. A government

[23] Kunzru, *Public Services in India*, p. 148. Recruitment of engineers for the state railways was combined with the selection process for the PWD, though relatively few engineers were recommended for the railways each year. Selection Committee reports for various years in 'Regulations Governing Appointments of Engineers' [APAC: IOR/L/PWD/5/29].

[24] 'Regulations as to Appointments of Assistant Engineers, 1910', *India Office List for 1910*, p. 222.

[25] Reply of the Master of Elibank, Under-Secretary for India, to Dr Rutherford, House of Commons Debate, 30 September 1909, vol. 11, cc. 1421–2 (Hansard): 'Indian Public Works Department (Assistant Engineers)'. This and all British parliamentary debates hereinafter cited (unless otherwise specified) were accessed via http://hansard. millbanksystems.com/.

[26] Selection Committee Reports for various years, in 'Regulations Governing Appointments of Engineers' [APAC: IOR/L/PWD/5/29].

[27] Selection Committee Report for 1912, in 'Regulations Governing Appointments of Engineers' [APAC: IOR/L/PWD/5/29].

[28] See Annexure XVIII, *Islington Commission Report*, p. 328 (paragraph 8).

report showed that in 1913, statutory Indians constituted 18.5 per cent of posts in the PWD with a salary of Rs 500 per month and above. As engineers below this salary level were more likely to be members of the Provincial Service than of the Imperial Service, this calculation probably underestimates the percentage of Indians. Yet, even accounting for this, it seems unlikely that the strength of the Provincial Service (to which most Indian engineers belonged) constituted 30 per cent of PWD engineers at this stage.[29] There was a temporary break in the recruitment of European engineers during World War I, when military service was made compulsory in Britain. Even this, however, did not affect the proportions in the long run: the shortfall was made up in 1919, when seventy-five vacancies were created (for appointments by the Secretary of State) for European candidates who had served in the war or 'been prevented on adequate grounds from so serving'.[30]

Reorganization of the Engineering Services, c. 1920–40

In the interwar period, the structure of the engineering services underwent several changes. Three factors drove this process. First, the difference in prestige and terms of employment between Imperial and Provincial engineers caused dissatisfaction in the latter group. Second, there was a growing political demand for Indianization in the public services, which was very relevant to the engineering services given their low percentage of Indians. Third, the constitutional reforms of 1919 and 1935 altered the relationship between the PWD and the central and provincial governments, necessitating a corresponding restructuring of the engineering services.

Salaries were an open secret in the Indian services. Kipling, in one of his vivid literary paeans to the colonial officer, explains a Punjab

[29] See table in *Islington Commission Report*, p. 25. The figure of 18.5 per cent is calculated for 'Indian and Burmans' and 'Anglo-Indians' combined. On the salaries, see *Historical Retrospect of Conditions of Service in the Indian Public Works Department (All India Service of Engineers)*, Private pamphlet (c. 1925) in the Secretary of State's Library Pamphlets, vol. 72, T.724 [APAC: P/T724], p. 8.

[30] Edwin S. Montagu (Secretary of State) to the Governor General of India ('Public Works, No. 6'), 6 February 1919, in 'Regulations Governing Appointments of Engineers' [APAC: IOR/L/PWD/5/29].

policeman's shabby home with the remark that 'in a land where each man's pay, age, and position are printed in a book, that all may read, it is hardly worth while to play at pretences in word or deed'.[31] Perhaps not, but the Provincial Service engineers certainly thought it worthwhile to protest against the distinctions in salary within the PWD: they received roughly two-thirds the pay of their Imperial counterparts (except at Chief Engineer rank, where salaries were equal).[32] This distinction between statutory Indians and Europeans was common to most services, and officials justified it on the grounds of 'the cost of production of a European officer in Europe'. They argued that 'no European will serve away from his own country without an exceptional inducement',[33] while paying Indians more than 'what is required to obtain suitable Indian officers' would 'impose for all time on the country a burden which she ought not to bear'.[34] The salaries (in rupees) of PWD engineers at various ranks until 1908 can be seen in Table 3.1.

Another source of discontent for the Provincial engineers was their status. As explained in Chapter 1, the term 'provincial' in most government services (such as the civil service) referred to officers who were given a lower grade of tasks, and could never attain the higher positions allotted to the Service. This association of the term with a lower

Table 3.1   Salary scales (in rupees per month) of PWD engineers, 1892–1908

Period	Type of service	Assistant Engineer	Executive Engineer	Superintending Engineer	Chief Engineer
1892–1908	Imperial	350–550	700–1,000	1,250–1,600	1,800–2,500
	Provincial	250–400	475–650	750–1,050	1,800–2,500

Source: Historical Retrospect of Conditions of Service in the Indian Public Works Department (All India Service of Engineers), private pamphlet (c. 1925) in the Secretary of State's Library Pamphlets 72, T.724 (available in APAC: P/T724), p. 8.

[31] Rudyard Kipling, 'William the Conqueror', in *The Second Penguin Book of English Short Stories* (London: Penguin, 2011 [1972]), pp. 61–94, here p. 64.

[32] *Historical Retrospect of Conditions of Service*, pp. 2, 8.

[33] *Islington Commission Report*, p. 38.

[34] *Islington Commission Report*, p. 37.

status affected the Provincial engineers too, although they performed the same range of functions as Imperial engineers, and indeed could attain the same ranks (including that of Chief Engineer).[35]

The Provincial Service of Engineers had originally been created for a limited period of time, which came to an end around 1906. Hopes were raised for a more equitable system, and Provincial engineers (or their representatives) submitted a petition to the government, asking for the Provincial Service to be dissolved. They were disappointed.[36] Members of Parliament sympathetic to the Indian/Provincial Service engineers raised questions about the unequal system in the British Parliament, but Secretary of State Morley refused to change the 'general rule of the public service in India that officers recruited in India, whether Indians or Europeans, are on a different footing as regards pay, leave, and pension from those who are recruited in England'.[37] In the reorganization that followed in 1908, the gulf in status between the Provincial and Imperial Services widened: different promotion criteria were created for the Provincial Service. Its members would have to serve for fifteen years as Assistant Engineers before receiving the rank of Executive Engineer; for Imperial engineers the period was just eight years.[38] Even as the Labour MP James O' Grady

[35] *Islington Commission Report*, p. 19 (paragraph 25). A few Provincial Service engineers had reached the highest ranks in 1912, when three Chief Engineers out of thirteen were alumni of the Indian engineering colleges (that is, members of the Provincial Service). The number for Superintending Engineers was twenty-one out of seventy-one. Calculated from Government of India, Public Works Department, *Classified List and Distribution of Establishment, corrected up to 30th June 1912* (Calcutta: Superintendent Government Printing, India, 1912 [APAC: IOR/V/13/216]). In this and all following statistics from *Classified Lists*, numbers for each rank include temporary or officiating rank holders.

[36] Kunzru, *Public Services in India*, p. 147.

[37] Exchange between Thomas Hart-Davies and Secretary of State Morley, House of Commons Debate, 10 February 1908, vol. 183, c. 1391 (Hansard): 'Appointments for Natives in Public Works Department, India'. Hart-Davies was Liberal MP for Hackney North. 'Members', House of Commons Parliamentary Papers, available at http://parlipapers.chadwyck.co.uk/fullrec/members.do?member=04266 (accessed 14 July 2012).

[38] Kunzru, *Public Services in India*, p. 147; *Historical Retrospect of Conditions of Service*, p. 3; Exchange between Thomas Hart-Davies and Thomas Buchanan, House of Commons Debate, 1 March 1909, vol. 1, cc. 1225-7W (Hansard): 'Indian Colleges (Imperial Service)'.

suggested in 1909 that the Imperial/Provincial distinction be done away with altogether,[39] Provincial engineers kept up their protests. In 1912 the old system—whereby their normal rate of promotion was the same as that of Imperial engineers—was restored.[40]

Meanwhile, as we saw in Chapter 1, demands for the Indianization of the public services were gathering momentum. The central and provincial legislative councils were expanded and partially Indianized in 1909; and the efforts of Indian legislators led to the appointment of the Islington Commission in 1912 to look into the Indianization of the public services and the salary and benefits of public servants.[41] The Islington Commission's report, published in 1917, noted that despite the existence of two services in the PWD, 'the superior duties are performed by one class of superior officers, recruited on a single standard of qualifications'. Hence, the Commission recommended that the Imperial and Provincial Services should be merged. 'In this way we should achieve an organisation of the services based on the work which they are required to do, and not on the race of, or the salaries drawn by, their members or any such artificial distinction.'[42]

This resulted in the most significant change in the PWD's organization since the creation of the Provincial Service in the 1890s: in 1920 the Imperial and Provincial Services were replaced by a single Indian Service of Engineers (ISE).[43] Also on the recommendation of the Islington Commission, fresh recruitment to the ISE was in the proportion of half and half in England and India respectively. While statutory Indians could still be selected by the Secretary of State, the 10 per cent quota for them in London was removed.[44] This 50–50

---

[39] Exchange between Mr O'Grady and Mr Harold Baker, House of Commons Debate, 10 October 1912, vol. 42, cc.500-1 (Hansard): 'Provincial Engineers (India)'. James O'Grady was Labour member for Leeds East. 'Members', House of Commons Parliamentary Papers, available at http://parlipapers.chadwyck.co.uk/fullrec/members. do?member=05256 (accessed 14 July 2012).

[40] Kunzru, *Public Services in India*, p. 147; *Historical Retrospect of Conditions of Service*, p. 3.

[41] *Islington Commission Report*, p. 2.

[42] *Islington Commission Report*, p. 19 (paragraph 26).

[43] Mital, *History of the Thomason College of Engineering*, pp. 181, 183.

[44] Annexure XVIII, *Islington Commission Report*, p. 329.

formula constituted the Commission's recommendation towards Indianizing the PWD (they calculated that the earlier share of engineers recruited in India had been 37.5 per cent).[45] Engineers already serving were appointed retrospectively to the ISE. Pre-1920 Upper Subordinates could also be promoted to the ISE.[46] The strength of the newly created ISE may be estimated at about 700.[47]

Although the ISE had been created primarily to remove distinctions between engineers on the basis of race and place of recruitment, some persisted. Engineers from the Indian colleges were still identified as India-recruited ISE and were employed on different terms from the London recruits. For instance, engineers domiciled outside India (that is, most of those recruited in London) received an additional overseas allowance.[48]

These changes were accompanied by another one, which, while increasing opportunities for India-trained engineers, perpetuated the two-tier system of engineer officers in the PWD. A new provincial service of engineers was created *for each province*. The existing Upper/Lower Subordinate branches were merged into one subordinate branch, and a large number of the erstwhile Upper Subordinates now populated the new provincial service (along with some direct recruits from the Indian engineering colleges). These men now occupied the rank of Assistant Engineer, and were placed in charge of some subdivisions. The lowest rank in the ISE was renamed Assistant Executive Engineer.[49] This may be summarized as follows:

---

[45] Annexure XVIII, *Islington Commission Report*, p. 328.

[46] See Figure 3.2 on p. 111.

[47] The sanctioned strength of Imperial and Provincial Services combined in the PWD at the time of the Islington Commission (c. 1915) was 728 (Annexure XVIII, *Islington Commission Report*, p. 326). As explained later, steps were taken to reduce the number of engineers in the lowest rank of the ISE.

[48] PWD *Classified Lists*; *Historical Retrospect of Conditions of Service*, p. 4; see also *East India (Civil Services in India): Report of the Royal Commission on the Superior Civil Services in India* (Cmd. 2128, London: HMSO, 1924), pp. 30–1 on overseas pay for officers domiciled outside India. The last of these sources is hereinafter cited as *Lee Commission Report*.

[49] Mital, *History of the Thomason College*, p. 183; PWD *Classified Lists*.

110   THE BIRTH OF AN INDIAN PROFESSION

ISE	
Chief Engineer	(manages province posted to)
Superintending Engineer	(circle)
Executive Engineer	(division)
Assistant Executive Engineer	(sub-division)
New provincial service (in each province)	
Assistant Engineer	(sub-division)

In proposing this change, the Islington Commission had reasoned that engineers entering the upper service (now the ISE) should not be kept at the lowest rank for an unduly long period, lest they should get bogged down in mundane tasks that did not make full use of their abilities. By converting the Upper Subordinate branch into a lower engineering service (with a corresponding increase in pay) whose members could be placed in charge of a certain number of subdivisions, the number of engineers in the lowest rank of the ISE could be reduced, allowing for quicker promotions.[50]

The new provincial services were large, with a cumulative strength of 850 at inception.[51] The Assistant Engineers who made up these services were mostly former Upper Subordinates or fresh graduates from the Indian colleges, which meant that the majority of them were statutory Indians.[52] These Assistant Engineers also had the prospect of promotion into the ISE ranks. The various paths into the ISE, through promotion and direct recruitment, are summarized in Figure 3.2.

Overall, there were increased opportunities for the graduates of the Indian engineering colleges. The patronage for Roorkee, for instance, was increased, guaranteeing appointments to the top nine or ten

---

[50] Annexure XVIII, *Islington Commission Report*, pp. 327–8.

[51] Mital, *History of the Thomason College*, p. 183. This number is commensurate with official data for the Madras Provincial Service in 1922, which had 100 Assistant Engineers. Government of Madras, PWD, *Classified List and Distribution Return of Establishment, corrected up to 30th June 1922* (Madras: Superintendent, Government Press, 1922 [APAC: IOR/V/13/831]).

[52] This is confirmed by the Madras list for 1922, in which 85 out of 100 Assistant Engineers bear recognisably Indian names—the remaining most likely Anglo-Indians and Domiciled Europeans. *Classified List and Distribution Return of Establishment* [APAC: IOR/V/13/831].

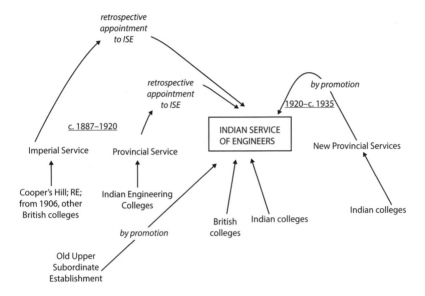

Figure 3.2 Routes into the Indian Service of Engineers (ISE) (direct recruitment and promotion)

Source: Compiled by the author on the basis of official PWD *Classified Lists* for various years.

graduates of its engineering class annually in the ISE.[53] While even more opportunities were created for graduates of the Indian colleges in the new provincial services, it is arguable that the new Assistant Engineer position was essentially a renamed subordinate position (with some increase in pay and responsibility, but lower status than the ISE ranks).

Despite the increase in prospects for Indian engineers, Indian opinion was sharply critical of the Islington Commission's approach to Indianization. One commentator felt that while the system of 50–50 recruitment was an improvement on the earlier opportunities for Indians, the Commission had not adequately justified its premise that recruitment in Britain could not be abolished entirely. He quoted the evidence of the PWD Secretary to the Government of India, who had said that 'no political considerations were involved' in selecting engineers for the department, and that the best trained engineers

---

[53] This guarantee to Roorkee continued until 1927. Mital, *History of the Thomason College*, p. 188.

should be selected no matter where they came from; the Indian engineering colleges, on the other hand, were accepted, even by the Commission itself, as of sufficiently high standard. Such being the case, the Commission's vague statement that they were 'satisfied that there are grounds of policy for continuing to recruit from Europe as well as India' was unsatisfactory.[54] As a matter of fact, it was becoming difficult to attract European candidates in Britain for the PWD; but it was still considered essential to recruit them.[55]

The percentage of Indians in the ISE did increase under the 50–50 recruitment policy, albeit gradually—as one would expect, given the large size of the service relative to the annual number of recruits. In 1924, when it had been in operation for a few years, the overall proportion of Indians in the ISE was less than 30 per cent.[56] Indian politicians' dissatisfaction with the rate of Indianization in the public services after World War I was noted, and the Lee Commission of 1923–4 was asked, among other things, to suggest measures for further Indianization.[57]

Provincialization

Alongside its measures on Indianization (discussed later), the Lee Commission also recommended an important change in the structure of the PWD's engineering services. This was made necessary by the introduction of dyarchy under the 1919 Government of India Act. As mentioned in Chapter 1, Roads and Buildings had now become a 'transferred' subject under the control of elected provincial ministers. To reflect this change, the Commission provincialized the corresponding branch of the ISE. This meant that no fresh recruitment would be

---

[54] Kunzru, *Public Services in India*, pp. 148–50. The Islington Commission's words quoted here are from Annexure XVIII of its report, p. 328.

[55] See Secretary of State's (Lord Birkenhead) letter ('Services, No. 27') to the Governor General of India dated 21 June 1928 in 'Regulations Governing Appointments of Engineers' [APAC: IOR/L/PWD/5/29], and enclosures thereto.

[56] *India Office List* for 1924. This estimate may be slightly on the lower side, as only officers drawing Rs 700 a month or more are listed, which means a number of Assistant Executive Engineers—half of whom were recruited in India—may not be accounted for.

[57] *Lee Commission Report*, pp. 5–6.

carried out for the Roads and Buildings branch of the ISE (although the existing officers would continue): these posts would be filled, as vacancies arose, by the concerned provincial governments. Indianization in this branch was therefore expected to proceed automatically, as the provincial governments would ordinarily recruit statutory Indians.[58]

For the Irrigation branch of the ISE, recruitment by the Secretary of State continued. The Lee Commission recommended that engineers for this branch be recruited in the ratio of 40 per cent Europeans, 40 per cent Indians, and 20 per cent Indians promoted from the provincial service. In the smaller provinces where the PWD was not divided into branches, the existing system of 50 per cent recruitment in England and 50 per cent in India was to be continued, while the old 10 per cent rule was reinstated (that is, of those recruited in England, 10 per cent must be Indians).[59]

Over the following years, the provincialization of Roads and Buildings led to a slow decrease in the size of the ISE, while the Indian quota in recruiting engineers for the Irrigation branch led to a gradual Indianization of the Service. In 1928, Indians formed around 36 per cent of the ISE officers included in the *India Office List*.[60] The process was accelerated when, following the Government of India Act of 1935, the PWD's Irrigation branch was also placed under elected ministers in the provinces, and fresh irrigation engineers recruited there rather than through the ISE.[61] Yet, British parliamentarians resisted this move, and the Secretary of State reserved the right to appoint, in exceptional circumstances, anyone of his choosing to any public post related to Irrigation.[62]

---

[58] In case European officers were required to be recruited in Britain for a provincialized service like Roads and Buildings, such recruitment would be carried out not by the Secretary of State but by the High Commissioner for India. See the reply of Earl Winterton, Under-Secretary of State for India, to Mr Wallhead, in 'Head [sic] Commissioner for India', House of Commons Debate, 29 March 1926, vol. 193, cc. 1624–6 (Hansard), here c. 1625.

[59] *Lee Commission Report*, pp. 21–2 (paragraph 40), 65 (paragraph xvi).

[60] Calculated from *India Office List* for 1928.

[61] That is, all fresh recruitment for the ISE would now cease. See J. D. Shukla, *Indianisation of All-India Services and Its Impact on Administration* (New Delhi: Allied Publishers, 1982), p. 335. See also Chapter 1.

[62] *Government of India Act, 1935* (26 Geo. 5, Ch. 2), p. 149 (paragraph 245). The debate on provincializing the Irrigation branch is discussed in detail in the following section.

With the PWD now entirely provincialized, the ISE's strength, as counted from the *India Office List*, was down to 369 (about half its original size) by 1940. Of these, 209, or 57 per cent, were Indians.[63] Indianization had progressed considerably from the figure of 30 per cent a decade and a half earlier. This should be seen in the context of two contemporary factors which were outlined in Chapter 1. The first was that from around 1920, it was increasingly difficult to attract British recruits to the public services as a whole. Second, as the Lee Commission acknowledged, demands were increasingly being made in the Legislative Assembly and elsewhere for a more emphatic move towards Indianization, including, for instance, transferring control of the All-India Services from the Secretary of State to the Government of India.[64] Despite this combination of factors, demands for far-reaching measures—such as recruiting only in India—were consistently held off until the 1935 Act came into force (and even then significant caveats were made). The government embarked upon Indianization and provincialization with considerable circumspection. The reasons for this were never declared overtly. Yet, as the following sections illustrate, one can discern some of the reasons by analysing the culture of engineering in the PWD, and the characteristics that British officials and policymakers thought the ideal engineer should possess.

THE CULTURE OF PUBLIC WORKS ENGINEERING

The Engineer as Gentleman and Generalist

From c. 1870 onwards, the India Office cast its India-bound engineers in a gentlemanly mould. As outlined earlier, applicants to the RIEC at Cooper's Hill were tested in classical languages and English history in addition to mathematics and natural sciences. Once admitted, students worked hard but also had a lifestyle involving billiards, wine, and formal meals in Hall.[65] Colonel Chesney, the first President of Cooper's Hill, was of the opinion that men selected to serve in the Indian PWD should be 'not only good engineers but religious

---

[63] Calculated from *India Office List* for 1940.

[64] *Lee Commission Report*, p. 6; Reginald Coupland's Minute in *Lee Commission Report*, pp. 116–23, here p. 120 (paragraph 12).

[65] Cuddy, 'Royal Indian Engineering College', pp. 171–2.

men, at any rate Christians in feeling and profession',[66] and in 1878 a military officer in the North Bengal State Railway expressed the view 'that a better, abler and more gentlemanly set of men than the recently joined men from Coopers Hill could not have been sent out to India'.[67] Indeed, these requirements were similar to the general characteristics prized by British officials in charge of selecting other types of colonial civil servant, such as members of the Indian Civil Service and its analogues in Malaya and Africa. As historians have shown, they valued pursuits such as riding and sport in candidates above mere intellectual prowess. Colonial service aspirants could be rejected for speaking with a Birmingham accent or selected on the strength of having been a Rugby Blue.[68]

Graduates of Cooper's Hill, who liked to talk of the '*esprit de corps*' in the Imperial Service that resulted from their years together in College,[69] had a formative influence on the culture of the PWD. Although the College was closed in 1906, its alumni dominated the upper echelons of the engineering services through to the 1930s, when the last batch of RIEC graduates came to the end of their careers. In 1912, RIEC alumni accounted for nine out of thirteen Chief Engineers (69 per cent) and forty-five out of seventy-one Superintending Engineers (64 per cent), figures in both cases including Imperial and Provincial Services,[70] and occupied virtually every position in the Government of India's PWD Secretariat.[71] Ten years later, when the ISE had

[66] Quoted in Cuddy, 'Royal Indian Engineering College', p. 200.

[67] J. G. P. Cameron, *A Short History of the Royal Indian Engineering College: Coopers Hill* (n.p.: Coopers Hill Society, private circulation, 1960), p. 11.

[68] Maria Misra, 'Colonial Officers and Gentlemen', pp. 153–4 and *passim*. David C. Potter identifies generalism and gentlemanliness as part of the culture of the Indian Civil Service (and later the Indian Administrative Service). Potter, *India's Political Administrators*, p. 247.

[69] Annexure XVIII: 'Public Works and Railway (Engineering) Departments', *Islington Commission Report*, p. 329 (paragraph 11).

[70] Calculated from Government of India, PWD, *Classified List and Distribution Return of Establishment, corrected up to 30th June 1912* (Calcutta: Superintendent Government Printing, India, 1912 [APAC: IOR/V/13/216]). Numbers for each rank include engineers holding temporary or officiating rank (for example, a Superintending Engineer who is also an officiating Chief Engineer is counted under both heads.)

[71] Government of India, PWD, *Classified List and Distribution Return of Establishment, corrected up to 30th June 1912.* The positions were—Secretary to

been formed, the RIEC's share of Chief Engineers (twenty-one out of thirty-four, or 62 per cent) and Superintending Engineers (fifty-seven out of 109, or 53 per cent) had dipped slightly, but was still considerable.[72] Consequently, the requirement for gentlemanliness outlived the College itself, as I will show shortly.

There was a second important requirement of the PWD engineer: that he be a competent generalist, able to tackle a wide range of problems. The maintenance of the physical infrastructure of government and the state, and of the canal and power systems sustaining agriculture, required a diverse set of skills. In 1870 the *Spectator* of London had described the qualities required of the engineer working in India thus:

> The ideal Engineer for India is a man who will take £1000 a year as his average income for life, and insist that all under him shall be content with their wages; who can build anything from a Tanjore tank as big as the lake of Lucerne to a cloacae for the last new stockade; who will regard an offer of a commission from sub-contractors as a deadly insult; who can keep accounts like a bank clerk... [73]

More than thirty-five years later the India Office continued, in selecting engineers for the PWD, to prize a strong grounding in a range of areas, supplemented by practical experience. After the closure of the RIEC in 1906, engineering graduates appearing for interview at the India Office were advised to prepare the following imposing list of subjects, considered important for service in the Department:

> *Pure Mathematics*, including a knowledge of the differential and integral calculus.
> *Applied Mathematics.*
> *Geometrical and Engineering Drawing.*
> *Surveying and Geodesy.*

Government: W. B. Gordon; Deputy Secretary: G. H. le Maistre; Under-Secretary: P. Hawkins; Inspector-General of Irrigation: M. Nethersole.

[72] Calculated from: Government of India, PWD, *Classified List of Establishment, corrected up to 30th June 1922* (Calcutta: Superintendent Government Printing, India, 1922) [APAC: IOR/V/13/220].

[73] Quoted in Black, 'The Military Influence', p. 232.

Strength of Materials and Theory of Structures.

Hydraulics.

Heat Engines.

Materials used in Construction.

Building Construction.—Wood and metal work, limes and cements, and building with stone, brick, and concrete.

Knowledge of the principles of road-making, waterworks, sanitary and railway engineering. (Important.)[74]

The 'gentleman generalist' paradigm of public works engineering is clearly illustrated in the typescript memoirs of a PWD engineer, G. F. Hall. After graduating from the Central Technical College in London (c. 1909), Hall applied for a position in the Indian PWD and was interviewed at the India Office. Prior to the interview, he had 'spent weeks mugging up engineering formulae, details of cement and brick manufacture and the weights and composition of materials'. On the actual day, he was asked what his best sport had been at school in Marlborough (he played cricket, football, and hockey, but was not exceptionally good at any of them); how he spent his leisure hours when at the Central Technical College (he said he worked all the time, but was pressed into admitting that he spent his leisure time on football, rowing, and dancing); how he found Chatham, where he was undergoing a short training course as a member of the Special Reserve of the Royal Engineers (he liked it very much). '[A]nd so much for all my weeks of cramming!'[75] Hall's social and sporting accomplishments and his institutional pedigree clearly carried as much weight with the committee as his technical qualifications.

Hall was placed on a waiting list following his interview, but received an appointment the following year as Assistant Engineer (Imperial Service of Engineers) in the PWD, where he soon had to prove himself as a generalist.[76] Shortly after his arrival in India in

[74] 'Note for the Information and Guidance of Candidates', *India Office List* for 1921, p. 254.

[75] G. F. Hall, 'All in the Day's Work' (typescript, 2 vols, 1947 [APAC: MSS Eur D569/1 and D569/2]), here vol. 1, p. 43 (on the interview), p. 41 (on the Chatham course), and p. 40 (for Hall's engineering qualifications).

[76] The Record of Services in the *India Office List* for 1940 gives the following details: Hall, Geoffrey Fowler, CIE, MC, Indian Service of Engineers, Superintending Engineer,

1911, he was placed in charge of a subdivision on the Tribini canal in northern Bihar. He continued in irrigation work, interrupted by war service in Greece and France from 1916 to 1918. Some years after he returned to India, he was transferred to the Roads and Buildings branch of the Bihar and Orissa PWD. Although he had little experience of building work and had failed an exam on building subjects early in his career, Hall was now given several assignments including the construction of schools, hostels, and barracks, and, in 1928, a host of arches and pylons as road decorations during the visit of the Viceroy and the Simon Commission to Patna.[77]

At every rank an engineer's duties were many and varied. As an Assistant Engineer, he had to negotiate contracts, establish rates, assess finished work, and draw up bills. An official report claimed that in some subdivisions, where the work was mainly related to maintenance, 'the duties of an Assistant Engineer make little demand on engineering skill'.[78] A similar view was apparently held by Frank Harris, a Cooper's Hill alumnus and Executive Engineer in the Punjab, who 'held the distinction of submitting a memorial annually to the Secretary of State for India.... He prayed that he should be permitted to retire as he found he was utilized, not as an Engineer but as a mere cooly-driver'.[79] At other times the Executive Engineer might also find himself 'undertaker to the Christian community'.[80] A Superintending Engineer was as much an administrator as a technical officer. His duties included supervising budgetary and accounts issues of his department, seeking and dispensing advice on technical matters,

---

Bihar. Born on 9 March 1888. Joined the service 1 October 1911. Assistant Engineer: November 1911; Executive Engineer: October 1920; Officiating Superintending Engineer, May 1931 and again February 1933; confirmed, March 1933; Chief Engineer (temporary), January 1936.

[77] Hall, 'All in the Day's Work', vol. 1, *passim*. The Simon Commission, controversial because it had no Indian members, was set up to assess the effects of the Montagu–Chelmsford Reforms of 1919. John F. Riddick, *The History of British India: A Chronology* (Westport, CT: Praeger, 2006), pp. 106–7.

[78] Annexure XVIII, *Islington Commission Report*, p. 327.

[79] G. E. C. Wakefield, *Recollections: 50 Years in the Service of India* (Lahore: Civil and Military Gazette Ltd., 1942), pp. 27–8. Courtesy of the Centre of South Asian Studies Archives, Cambridge.

[80] Hall, 'All in the Day's Work', vol. 1, p. 133.

managing subordinate personnel, and corresponding with the provincial government (see the list provided later). As Chief Engineer
the PWD officer was responsible for preparing annual reports on the
activities of his department.[81] Engineers could also be deputed for a
period of time on 'foreign service' to another province or department;
for instance, some ISE officers of the United Provinces PWD were
deputed to the Department of Education and thence as professors
to Roorkee in 1934–5.[82] They were also frequently seconded to the
service of princely states for brief periods.[83] The diversity of a PWD
engineer's duties is vividly illustrated by the following list of the files
on G. F. Hall's desk on a single day in 1933, when he was a Superintending Engineer:

Demarcation of Provincial roads.

Furniture in Circuit Houses.

Failure of District Board causeways.

Chief Engineer's Inspection Note.

Ventilation in Police Barracks.

Encroachments.

Progress of experiments. Chief Engineer's comments.

Accountant-General's objection items in Divisional accounts.

Retired clerk wants reemployment.

Allotment of Deputy Magistrates' Quarters.

Estimator wants extension of leave.

Transfer of Police buildings from District Board to P.W.D.

Quadrennial repair estimates and allotments.

Additions and alterations to new Police Lines at Chaibassa.

Recovery of rent from the Headmaster of a school.

Disposal of an old mortuary.

Major works must have a budget provision.

Dismantlement of a Police latrine.

[81] For example, *Administration and Progress Report of the Chief Engineer, United
Provinces Public Works Department, Buildings and Roads Branch, For the year 1934–35*
(Allahabad: Superintendent, Printing and Stationery, United Provinces, 1936),
addressed by Chief Engineer Chhuttan Lal, to the Secretary to Government (United
Provinces PWD, Roads & Buildings [APAC: IOR/V/24/3304]).

[82] *Administration and Progress Report of the Chief Engineer United Provinces Public
Works Department, Buildings and Roads Branch, For the year 1934–35*, chapter 1, p. 2.

[83] See, for instance, 'R.J.I.', letter to the editor, *Indian Engineering*, 1 January 1887,
pp. 4–5.

Site plan required for completion of Site Committee's proceedings.

A long list of road metal rates for approval.

Dismantlement of P.W.D. Inspection Bungalows no longer required.

Executive Engineer cancels casual leave.

Repair of machinery.

Special repair of machinery.

Metal collection and consolidation estimate of a road for sanction.

Charges for sanitary services in Circuit Houses.

Recoveries for materials issued to a contractor.

Revision of Municipal taxes.

An Executive Engineer wants permission to avail himself of the Pujah holidays.

Reminder to Government for certain land plans.

Petition for reinstatement from dismissed work-charged establishment.

Contract for repair of typewriters.

Big bridge design. Steel or submersible in reinforced concrete?

Grants required by District Boards.

District Board seeks advice re causeways.

Revision of standard measurements.

Advisability of deepening wells sunk through rock.

Inspection note on District Engineer's office.

Inspection note on an S.D.O.'s office.

Subsidence of roads in the mining area.

Candidates for P.W.D. Professional Examination.

Postings due to leave.

Leave without pay for not following medical advice.

Minor Works Return.

Transfer of Overseers.

Stocking lakes for fishing leases.

Appeals from contractors.

Application for post of 2nd. clerk.

A new building material from waste materials.

Short issue and recovery of materials.

Travelling allowance bills.

Many petitions.

Revision of building rates.

Idol and encroachment in Tasildhar's [sic] compound.

Allotment of funds from S.E.'s reserve.[84]

---

[84] Appendix I: 'Files on My Table When I Went to Office on September 7th 1933', Hall, 'All in the Day's Work', vol. 2, p. 550.

An engineer was supposed to learn on the job, by doing and by seeking help where necessary. In 1930 Hall was asked to prepare a design for a bridge over the river Son at Dehri in Bihar. Dehri had a railway bridge belonging to the East Indian Railway (EIR), but no road bridge. Hall consulted irrigation engineers, the EIR's bridge engineer, and specialists working for European engineering firms in Calcutta before he drew up the plans. Recounting this incident in his memoir, Hall reflected that PWD engineers had to be resourceful, for they '[were] not specialists in any branch of engineering'.[85] Another skill a PWD engineer needed to develop was the ability to work in partnership with commercial contractors and suppliers of equipment and material. The link between British commercial interests and public works in India was strong: the British engineering firms of Bombay and Calcutta existed primarily to procure equipment and undertake contracts for the building of public works and the railways.[86] Soon after he had handed in his designs for the Dehri Bridge, Hall was summoned to Ranchi to help construct a jail for 2,500 prisoners, as demonstrators courted arrest *en masse* as part of the civil disobedience movement. Having drawn up a rough design with his colleague Ian MacRae, Hall engaged an engineer from a private firm, the Kumardhubi Engineering Works, to prepare detailed designs for a number of sheds. (In view of the urgency of the situation, Hall had free rein. In any case a Superintending Engineer could exercise his discretion in selecting a contractor.) Together they worked out the estimated costs before Hall travelled to Calcutta to formalize the contract with the Kumardhubi Works' managing agents, Bird & Co. He spent several days in Calcutta to order materials from the engineering firms there. The firms, scenting a big customer in the midst of the Depression, plied him with cocktails and free transport, while he negotiated orders at the upmarket restaurant, Firpo's, and the Barrackpore Golf

[85] Hall, 'All in the Day's Work', vol. 1, p. 248.

[86] Many engineering firms with European owners operated at this time, especially in Bengal. An example is the Gariahat Engineering Works, Ballygunge, proprietors TE Thomas and Co. Ltd, Calcutta. In 1930 they were able to provide sheet iron workers, oxy-acetylene welders, mechanical engineers; manufacture wrought iron railings and gates for government use; act as contractors to government, the PWD, railways, and municipalities. See entry in Commercial Industries section, *Thacker's Indian Directory* for 1930.

Course. Throughout his narration of this episode in his memoir, Hall presents himself as a heroic upholder of law and order, and especially as a tough negotiator and manager of men and materials, rather than as the creator of a technically challenging design.[87]

For young British engineers freshly arrived in India, their new environs could provide a heady mix of adventure and a sense of power. When Hall took up his first sub-division in January 1912, he found that his nearest colleagues were stationed tens of miles down the canal system. On his own not inconsiderable patch, Hall felt invincible:

> [W]ith a wife, a gun, two ponies, a trap, a car, a salary of Rs 380 a month and a Sub-Division of my own, I regarded myself as the equal of Kings and the ruler of the most beautiful kingdom [on] earth ... notwithstanding [the] difficulties [language], I entered into contracts, got along with my job and thoroughly enjoyed every minute of our jungle *dominion*.[88]

Some fifteen years later, another young British engineer, Herbert Fagent Merrington (b. 1904) was appointed to the ISE and posted in the Punjab Irrigation Service. Schooled in Surrey and trained in Guildford Technical School and Northampton Polytechnic, Merrington was in his early twenties when he was assigned to the Sutlej Valley Project, as part of which weirs were built at Suleimanki, Islam, Panjnad, and Ferozepur.[89] Arriving at the Islam Weir in the princely state of Bahawalpur (located in the Punjab province), Merrington found it 'a devil of a place.... Thirteen miles to the nearest railway station, a hundred miles from the nearest town, and two other white men besides myself. Thousands of natives & coolies'.[90] He may have been out in the country, but, as he assured his family in Surrey, 'I haven't grown a beard: we shave every day, and always put on a decent suit and collar & tie for dinner every evening'.[91] In fact the weir, which was nearly finished, had been

---

[87] Hall, 'All in the Day's Work', vol. 1, pp. 251–5.

[88] Hall, 'All in the Day's Work', vol. 1, p. 55. Emphasis mine.

[89] Biographical details in handwritten sheet in the Herbert Fagent Merrington Papers, European Manuscripts, British Library, London (MSS Eur D1229/1 and D1229/2, here D1229/1).

[90] Merrington to his family, 18 October 1927 (MSS Eur D1229/2, Sheet 6).

[91] Merrington to his family, 5 December 1927 (MSS Eur D1229/2; Sheets 12–15).

inaugurated by the Viceroy days after Merrington's arrival. 'I have an invitation [to the opening] of course,' Merrington wrote home; he had also been invited to the lavish breakfast that the Nawab of Bahawalpur was hosting for the Viceroy. 'Some binge! I think it's a good show having breakfast with the Viceroy about a fortnight after landing!'[92]

Shortly thereafter, Merrington was working on the construction of the Panjnad Weir, when he produced a vivid (if idealized) description of an engineer's daily work on the project. 'Dawn is breaking in purple-grey glory across the Bahawalpur desert', it begins, and goes on to detail the various tasks awaiting the engineer. He is woken by bearers before going out to the weir, about two miles from the engineers' bungalow, 'on his trolley pushed by four coolies'. At the site of the work there are eight hundred labourers; the engineer questions 'his subordinate, always cheerful and anxious to please', and carries out inspections before dealing with his mail in the office and returning to his bungalow for lunch. In the afternoon he returns to the works, 'this time wearing goggles and ... topee', braving the heat and dust. In the evening the engineer might preside as the workers are paid their monthly wages, then take up 'one or two other matters of importance ... with contractors, and another visit to the scene of work follows. The progress is not satisfactory—more stone must be brought during the night, and more men must be put on this section. Assurances are given by the subordinate that these orders will be carried out'. At long last the engineer is home for dinner, followed by 'a short spell of letter-writing, reading, or gramophone entertainment, and so to bed'.[93] Engineers new to India, in their early to mid-twenties, rapidly acquired the functions and bearing of the powerful administrator, whose accomplishments must range from dining with Viceroys to commanding legions of workers.

The Status of the Indian Engineer

The experiences of these expatriate engineers contrasted considerably with that of their Indian colleagues in the PWD. Most Indian

[92] Merrington to his family, 18 October 1927.

[93] Photocopy of typescript dated April 1928, Merrington Papers (MSS Eur D1229/1). The document begins: 'Dawn is breaking in purple-grey glory across the Bahawalpur desert ... '

officers were recruited from the Indian engineering colleges.[94] Notable among them were the alumni of the oldest of these institutions,
the Thomason College at Roorkee. Located in the United Provinces
(UP), home to the nineteenth-century Ganges Canal Project, and not
far from the Punjab, the fertile land of five rivers, Roorkee was particularly well placed to supply engineers for the growing irrigation
works in India.

In the long nineteenth century, it was theoretically possible for
graduates of the Indian colleges to build relatively successful careers
in the PWD, rising through the ranks despite starting out in the
'inferior' Provincial Service. A number of Roorkee alumni did make
a name for themselves. These were often Domiciled Europeans or
Anglo-Indians.[95] The campus itself was segregated: the Engineering
and Upper Subordinate courses at Roorkee were divided into English
and Indian Classes. 'It is doubtful,' writes Roorkee's historian, 'if they
mixed anywhere.'[96]

The occasional 'Asiatic' Indian achieved prominence in the PWD,
but rarely attained the highest ranks in the service. Ganga Ram (b.
1851) joined the PWD as a Roorkee graduate in 1873, was associated
with the construction of several landmark buildings in Lahore, and
received honours from the colonial government including a Rai Bahadur and (in his later years) a knighthood. Yet it is telling that when he
retired in 1903, the highest rank he had attained was that of Executive Engineer (barring a brief spell as Superintendent of the Corona-
tion Durbar Works in Delhi in 1903). He subsequently served in the
higher rank of Superintending Engineer in the princely state of Pati-
ala. His real success, though, came when, upon retirement, he leased
20,000 hectares of barren land in the Montgomery District of the
Punjab, which he made productive by irrigating it using hydroelectric
power. The venture, which *Indian Engineering* called 'a great Indian
enterprise', brought Ganga Ram a fortune that helped him become

---

[94] A small number, as noted earlier, was recruited in Britain by the Secretary of
State.

[95] See the names mentioned in 'Roorkhee Then and Now', *Indian Engineering*, 13
March 1926, pp. 141–2.

[96] Mital, *History of the Thomason College*, p. 62.

a philanthropist.[97] Even Mokshagundam Visvesvaraya[98] (a graduate of the engineering college in Poona), who rose rapidly in the Bombay Irrigation Service, felt he had hit a glass ceiling in his late forties when he was Superintending Engineer. 'Remembering that there was political feeling in the country at the time,' Visvesvaraya recounted later, 'I thought there was little chance of Government appointing me Chief Engineer except when my regular turn came according to my regular rank.'[99] He quit the PWD, and soon after was appointed Chief Engineer in Mysore State. Upon his departure, the Governor of Bombay wrote him a warm valedictory note, remarking, 'I hope that you will feel on reflection that your experience in Government service up to the present time has been *exceptionally* fortunate'.[100] In later decades some Indians did reach the higher echelons of the bureaucracy, as in the case of A. V. Ramalinga Iyer and R. Swaminatha Iyer, who attained the rank of Chief Engineer in Madras in the 1920s;[101] Jwala Prasad, Chief Engineer Irrigation, United Province in 1932; and Madan Gopal Sardana, Superintending Engineer in the same department, and Principal of Roorkee, 1940–5.[102]

Most Indians, however, started modestly in the provincial services and dreamt of entering the ISE.[103] They had to be constantly on their

[97] Appendix to Mital, *History of the Thomason College*, p. 262; 'A Great Indian Enterprise', *Indian Engineering*, 28 March 1925, p. 175; 'Indian List. Knights', *The Aberdeen Daily Journal*, 3 June 1922, p. 8 (accessed via the online British Newspaper Archive); 'Gunga Ram', in *The Indian Biographical Dictionary 1915*, edited by C. Hayavadana Rao (Madras: Pillar & Co), p. 170; Ravi Nitesh, 'The Legacy of Sir Ganga Ram', *Daily Times* (Pakistan) online, 17 April 2014, available at http://www.dailytimes.com.pk/opinion/17-Apr-2014/the-legacy-of-sir-ganga-ram (accessed on 3 September 2015).

[98] 'Visvesvaraya, Sir Mokshagundam' in *Indian Biographical Dictionary 1915*, edited by Rao, p. 452.

[99] M. Visvesvaraya, *Memoirs of My Working Life* (Bangalore: M. Visvesvaraya, 1915), p. 32.

[100] Quoted in Visvesvaraya, *Memoirs of My Working Life*, p. 33. Emphasis mine.

[101] S. Muthiah, 'Engineers Who Made history', *The Hindu, Metro Plus*, 9 June 2008, online version, available at http://www.thehindu.com/todays-paper/tp-features/tp-metroplus/engineers-who-made-history/article1418243.ece (accessed on 3 September 2015).

[102] Appendix to Mital, *History of the Thomason College*, pp. 262–3, 265.

[103] For example, Ram Das of the Punjab PWD (Irrigation), who rose from sub-divisional officer to the ISE. His career is described in his son's memoir: Prakash

guard lest their competence or reliability should be questioned. As the son of one such engineer recalled:

> Father, who never relaxed from work, explained it to us by saying that Englishmen could afford to relax because if things went wrong they managed to explain it to each other, and took the attitude that things sometimes go wrong. But when an Indian made a mistake the reaction, *if an understanding one*, was that the job was perhaps too difficult for him; 'after all they did not have the skill or the experience; one must be careful with giving responsibility too soon....'[104]

British attitudes to these Indian engineers became particularly important in the interwar context of political reforms and Indianization. When senior European engineers and British policymakers pictured the ideal public works engineer, they thought not only of a generalist, a man of character and resource, but of a quintessentially *British* engineer. Indian engineers had to be made in their image, but could never quite hope to achieve the ideal of the gentleman generalist. For one thing, they were always in danger of committing a social faux pas. George Wakefield, a European irrigation engineer, narrated in his memoirs an incident from the 1890s, when he was serving in the Punjab. A Superintending Engineer had arrived for an inspection tour, along with his wife and four daughters.

> Several 'water babies', as we Irrigation Engineers were called, had collected for the inspection and one was an Indian, a very good fellow but also very green in Western ways. He enquired how many visiting cards he should send in when he called on the Superintending Engineer and his family. I chaffingly advised him to send in a whole packet as there were so many of them and let them take as many as they wanted. Dining with the visitors a few days later and sitting next to one of the daughters I was overwhelmed when she remarked, 'What a curious man Mr. ... is? Do you know that when he called upon us yesterday he sent in a whole packet of visiting cards?' I confessed and was forgiven.[105]

Tandon, *Punjabi Century: 1857–1947* (London: Chatto and Windus; Toronto: Clarke, Irwin and Co., 1961).

[104] Prakash Tandon, quoted in Klingensmith, *'One Valley and a Thousand'*, p. 232. Emphasis mine.

[105] Wakefield, *Recollections*, p. 27.

Some years later, G. F. Hall felt in a quandary when an Indian colleague was due to accompany him on an official tour of irrigation facilities in north-west India. The Indian's food habits were different from Hall's, who 'foresaw that we were unlikely to be entertained together, and that as I could not very well leave him to his own devices I should have to spend my time in dak bungalows and see nothing of the social life of the two Provinces'.[106]

British engineers' views of Indians ranged from the paternalistic to the overtly sceptical. Hall was never entirely at ease with his Indian colleagues and superiors, as revealed by his remarks about some of them in his memoirs. One Superintending Engineer, he claimed, would listen to the grievances only of his fellow caste men when on inspection tours.[107] When Hall was an Executive Engineer in the 1920s, he had an Indian Superintending Engineer, Rai Bahadur S. C. Chakarbatty, whom he thought 'a nice elderly Bengali but unwilling to accept responsibility or issue definite orders. He was consequently little more than a post office between myself and the Chief Engineer'.[108] Further on the theme of accountability, Hall recounted the case of an Indian engineer in 1934 who was transferred out of his Circle following the devastating Bihar earthquake of that year and replaced by a Briton. The engineer was not offended, but 'thanked [his superior officer] profusely for his kindness in relieving him of such responsibility!'[109] Hall's lukewarm attitude to his Indian colleagues was not due to any technical incompetence on their part. Instead, he thought they came up short on non-technical parameters of competence: integrity, courage, and the willingness to shoulder responsibility.[110]

---

[106] Hall, 'All in the Day's Work', vol. 1, p. 159.

[107] Hall, 'All in the Day's Work', vol. 1, p. 71.

[108] Hall, 'All in the Day's Work', vol. 1, p. 182.

[109] Hall, 'All in the Day's Work', vol. 2, p. 341.

[110] Ongoing research by Vanessa Caru on the PWD in Bombay Presidency also indicates the existence of a normative culture of gentlemanliness. Interestingly, Caru also argues that Indian engineers in the early twentieth century, in pressing their claims to being treated on a par with their European colleagues, 'promoted an alternative conception of the profession, inspired by liberal and nationalist ideas, based on merit, technical qualifications and the national interest of the work they undertook'. Nevertheless, she shows, the powers that be held on to their old ideals. Vanessa Caru, 'The Creation of a "Corps d'Etat"? Indian and European Engineers of the Bombay

Plain prejudice played its part too. An interwar enquiry into the declining popularity of service in India among British engineering graduates found that, among other things, they were anxious about Indianization: 'Graduates and students are unwilling to offer themselves for appointments in a Service which entails working under an Indian superior.'[111]

As we saw in the previous section, colonial officials and British legislators were also unconvinced of the suitability of Indian engineers for senior and responsible positions in the PWD, but found themselves walking a tightrope. On the one hand, even for those statesmen who would rather that the British did not loosen their control on governance, the interwar constitutional reforms were a *fait accompli* and a necessary evil, given official sanction by Secretaries of State and Viceroys, beginning with the pair of Montagu and Chelmsford in 1919. On the other hand, most policymakers were keen to ensure continuity in the government's bureaucracies and anxious not to diminish the attractiveness of these to existing and future British officers. In balancing the two considerations, they confronted and gave expression to their attitudes to Indian engineers, their technical competence, their administrative ability, and their general character.

In a Minute appended to the report of the Lee Commission in 1924, one of its members, Sir Reginald Craddock, a former Lieutenant-Governor of Burma, expressed his regret at the (inevitable) changes to the PWD's engineering services caused by the provincialization of its Roads and Buildings branch. Thereafter, recruits to the branch would almost certainly be Indians, of whom there were now enough with the required technical knowledge. '[W]ith such excellent engineering colleges as exist in India there will be no lack of qualified Indian candidates to construct and maintain such works as still remain in

Public Works Department (1860's–1940's)', presented at the mid-term workshop of the ENGIND (Engineers and Society in Colonial and Postcolonial India) project under the French National Research Agency, New Delhi, January 2016, p. 12. I thank Dr Caru for permission to quote from her work in progress.

[111] A. J. R. Hope to the India Office, 28 March 1928, p. 5. Hope's report is enclosed with the Secretary of State's (Lord Birkenhead) letter ('Services, No. 27') to the Governor General of India, dated 21 June 1928, in 'Regulations Governing Appointments of Engineers' [APAC: IOR/L/PWD/5/29].

provincial charge.' But technical competence was not enough. Crad-dock hoped that the British engineers currently in the Service would 'serve out their full time and impart their high standards of skill, duty and integrity to those who come after them'. He felt that 'it will be many years yet before [the Roads and Buildings branch of the ISE], with its high traditions, will disappear'. The 'high traditions', it would seem, were related to British character.[112]

The views of politicians of various hues found systematic expres-sion in a 1935 Parliamentary debate on the Government of India Bill. In relation to the proposed provincialization of the PWD's Irrigation branch, Patrick Donner, Conservative MP for Islington West, moved an Amendment seeking the continuation of recruitment by the Sec-retary of State as the primary means of staffing the Irrigation service, and was supported by several other Members of Parliament.[113]

Donner felt that the proposed provincialization of Irrigation 'will mean in practice the disappearance of the British element'. He argued that the 'lesson' from departments transferred earlier had been one 'of deterioration of efficiency and of administration'. Irriga-tion was essential to the welfare of India. If the Government ignored the importance of irrigation, they were 'risking ... the very existence of millions of the population, and put no greater value on these lives than they would on a 5-franc counter in the Casino at Monte Carlo'. Donner then linked the efficiency of irrigation work directly to the presence of British engineers. Irrigation in India was 'a purely Brit-ish creation *brought about by British integrity, resource and impartiality*, which has no counterpart in present times or during the centuries in the past'.[114]

[112] 'Minute by Sir Reginald Craddock upon Certain of the Conclusions of the Commission', *Lee Commission Report*, pp. 127–38, here p. 133.

[113] 'Clause 233. (Services recruited by Secretary of State.)', House of Commons Debate, 4 April 1935, vol. 300, cc. 582–635 (Hansard). Hereinafter cited as Hansard HC Debate 4 April 1935. The Amendment sought to retain recruitment by the Secretary of State for the Forestry Department as well as the Irrigation Department, but the debate was mostly about Irrigation. For his party/constituency details, see entry on Sir Patrick Donner in 'Members', House of Commons Parliamentary Papers, available at http://parlipapers.chadwyck.co.uk/fullrec/members.do?member=03756 (accessed 12 June 2012).

[114] Hansard HC Debate 4 April 1935, cc. 584–8. Emphasis mine.

The Duchess of Atholl (Conservative MP for Kinross and West Perthshire)[115] cited evidence given to the Simon Commission[116] on British and Indian irrigation engineers. Sir Charles Harrison, Chief Engineer of the large-scale Sukkur Barrage project in Sind (completed in 1932)[117] had 'stressed the tremendous importance of the impartial distribution of water'. He had told the Commission 'that Indian engineers, *though often technically efficient*, were subject to outside influences pressing on them to make a partial distribution of water, influences which often made it difficult for them to carry out their duties as efficiently and impartially as they would wish to do'. Harrison's 'own Indian officers had said to him that they themselves recognised that British officers were in a far stronger position than they were, because they were not exposed to the pressure of members of their community or their family to give them more than their fair share of water'.[118]

The question of impartiality was felt to be of paramount importance in the heavily agrarian provinces of Sind and (especially) Punjab. Here, one of the primary responsibilities of the irrigation engineer was the efficient distribution of canal water. 'In a dry country with scanty rainfall' such as West Punjab, receiving canal water was a matter of survival for the farmers—and if they did not get it, they 'could ... be very violent'.[119] The threat of violence and unrest in these

---

[115] 'Stewart-Murray, Katharine Marjory, Duchess of Atholl', *Concise Dictionary of National Biography*, Oxford University Press, March 1992 via http://www.knowuk. co.uk (accessed 29 January 2011).

[116] The Statutory Commission or Simon Commission (1927–30) assessed the effects of the Montagu–Chelmsford Reforms of 1919. Riddick, *The History of British India*, pp. 106–7.

[117] 'India's Great Barrage', *The Observer*, 17 January 1932, p. 11. Speaking on the opening of the barrage, the Governor of Sind, Lord Lloyd, said that although the current atmosphere was critical of British rule, and despite the fact that the project had had British engineers in senior and Indians in junior ranks, engineers of the two races had 'been co-operating quite willingly together'.

[118] Hansard HC Debate 4 April 1935, c. 594. Emphasis mine. That Indians felt pressurized by their own communities is corroborated by at least one independent source. Irrigation engineer Ram Das of Punjab, in his son's words, faced the 'difficulty of fitting a new definition of integrity into the traditional pattern of duties towards the family and caste. Father used to envy his English colleagues who, being far away from home, had no relations demanding favours'. Tandon, *Punjabi Century*, p. 34.

[119] Tandon, *Punjabi Century*, p. 50.

provinces was very much in the minds of the parliamentarians. Patrick Donner warned that if the land were to become unproductive, the people '[would] turn upon us in their rage and fury, because they will say that we are responsible for their poverty'.[120]

Communal pressures were one thing, but the supporters of the Amendment drew attention to an even bigger evil: corruption. Donner averred that 'bribery [was] going to enter into this question of water'.[121] The Duchess of Atholl referred to a Chief Engineer (Irrigation), United Provinces, who had told the Simon Commission that landowners and farmers bribed Indian subordinate officers (for larger shares of water), and that British engineers 'were more insistent on checking it' than Indian engineers.[122] Another MP, Vice-Admiral Taylor, expressed this view most emphatically. 'The whole basis of the irrigation service,' he said, was 'the efficiency, integrity, good administration, and freedom from the corruption and bribery which, unfortunately, undoubtedly exist in India as in other places . . . if we eliminate the English official from the irrigation service, there is no question that the administration of that service will go down.'[123]

Donner's supporters included Winston Churchill (Conservative MP for Epping),[124] who felt that 'this is one of the most frightful responsibilities that the House of Commons has ever been asked to take'. He painted a dire picture of Irrigation under Indian engineers after provincialization. Referring to the growing population of Punjab and Sind under artificial irrigation, he told the Secretary of State, Sir Samuel Hoare, that if recruitment was handed over to the provinces, he would be responsible for 'undermining the means by which these new-come millions get their new-found food'. He added that this was

[120] Hansard HC Debate 4 April 1935, c. 589.
[121] Hansard HC Debate 4 April 1935, c. 588.
[122] Hansard HC Debate 4 April 1935, c. 594.
[123] Hansard HC Debate 4 April 1935, cc. 615–16.
[124] Paul Addison, 'Churchill, Sir Winston Leonard Spencer (1874–1965)', *Oxford Dictionary of National Biography*, Oxford University Press, 2004; online edition, Jan 2011, available at http://www.oxforddnb.com/view/article/32413 (accessed 8 June 2012). Churchill was part of a 'vindicative [sic] majority' in the Parliament with whom 'a small band of members sympathetic to the Indian cause' engaged during the debates on the 1935 Government of India Bill as a whole. Burton Stein, *A History of India*, 2nd edn, edited by David Arnold (Oxford: Wiley-Blackwell, 2010), p. 326.

'only part of the general wreckage. It is only part of this vast process of liquidation of what Britain has done in India'.[125]

A few MPs disagreed. Edward Turnour, the Earl Winterton (Horsham; Conservative)[126] took exception to Churchill's remarks, which '[suggested] that [Indians] are so corrupt, so incompetent, so inefficient that if they are entrusted with this great charge they will undermine the whole fabric of India'. Winterton asked: 'Has ever a more serious charge been made against any people?'[127] Sir Murdoch MacDonald (Inverness; Liberal), a former irrigation engineer in Egypt and a recent President of the Institution of Civil Engineers,[128] felt that 'it would be very wrong to take out of the hands of the people of India the control of the irrigation service'. He did not think the issue of corruption was a real problem where there existed sufficient water. Corruption, if it did come into play, could only exist at the lowest subordinate levels, 'the people who turn the smallest valve'; from his experience in Egypt that type of corruption could not be stopped even by the presence of senior British engineers in the department.[129]

Macdonald's and Winterton's views notwithstanding, a great number of MPs had expressed a lack of faith in Indian engineers. As a concession to their views, the Secretary of State (Sir Samuel Hoare), representing the government, offered to 'consider the proposal that the Secretary of State should make recruitment of officers'. In the end Donner withdrew his Amendment on the basis of this 'assurance'.[130] Ultimately, the original compromise in the Government of India Bill was retained in the Act of 1935—recruitment for the Irrigation service would be carried out in India, with the Secretary of State retaining the option to appoint officers to any service or office concerned with

[125] Hansard HC Debate 4 April 1935, cc. 619–20.

[126] 'Turnour, Edward, sixth Earl Winterton and Baron Turnour', *Concise Dictionary of National Biography*, Oxford University Press, March 1992 via http://www.knowuk. co.uk (accessed 12 June 2012).

[127] Hansard HC Debate 4 April 1935, c. 621.

[128] H. E. Hurst, 'MacDonald, Sir Murdoch (1866–1957)', revised by Elizabeth Baigent, *Oxford Dictionary of National Biography*, Oxford University Press, 2004, available at http://www.oxforddnb.com/view/article/34709 (accessed 8 June 2012).

[129] Hansard HC Debate 4 April 1935, cc. 617–18.

[130] Hansard HC Debate 4 April 1935, c. 619 and c. 623.

irrigation in India if it was deemed necessary 'for ... securing efficiency in irrigation in any Province'.[131]

***

This chapter has offered the first full description of the changing
personnel, recruitment, organization, and administration of PWD
engineers over the period 1900–40. It reveals two important features
of the engineering services. The first is that despite the Government
of India's drive to Indianize its services, distinctions were always
made between European and Indian engineers. The separate Imperial Service and Provincial Service lasted until 1920; thereafter, within
the ISE, distinctions in salary and privileges continued to be made
between engineers of Indian and British domicile. The second important feature, which helps to account for this differential treatment, is
that many British officials and MPs viewed the European engineer
as a bulwark against the threat of lowered standards in the PWD—
which they thought would inevitably accompany Indianization. For
this reason, successive Royal Commissions rejected wholesale Indianization of the ISE, and British MPs resisted (though unsuccess-
fully) the proposed provincialization of the Irrigation branch of the
ISE. Such views held sway in spite of the difficulty of attracting British
recruits to India after World War I, and the acknowledged quality of
the Indian engineering colleges from which the Indian members of
the ISE were drawn.

Change did occur, albeit slowly: by the end of the 1930s, both
branches of the PWD had been placed under elected provincial ministers, the ISE was more than 50 per cent Indian, and recruitment in
Europe had all but ceased (although the Secretary of State retained
special powers to appoint irrigation officers). Crucially, though, this
change was carefully managed by the government, not only by maintaining a European core in the ISE, but by ensuring that the Secretary of State had control over it, and more generally over the PWD. I
have argued that the reasons for this are best understood by studying
the normative culture of public works engineering, as embodied in
the training, recruitment, and career paths of its practitioners. The

---

[131] *Government of India Act, 1935* (26 Geo. 5, Ch. 2), p. 149 (paragraph 245).

ideal of the public works engineer was cast in a British mould. He was to be a gentleman and a generalist, possessed of integrity, courage, and resourcefulness. British engineers and policymakers were unconvinced of the extent to which these qualities might be found in the Indian engineer. They held that while Indian PWD engineers might be technically competent, their personal qualities were likely to fall short of the ideal of the gentleman generalist, leaving them open to corruption, partiality, and incompetence as administrative officers. These ideas on race and technical expertise did not merely influence the experiences and relationships of PWD engineers and the ways in which the PWD was reorganized and Indianized; they also reflected, in microcosm, larger anxieties about the future of British rule in India and about Indians' ability to govern themselves.

CHAPTER FOUR

Keepers of the Peace

Efficiency, Loyalty, and the Limits of Indianization on the Railways

As employers of engineers in the period 1900–47, the subcontinent's railways were every bit as important as the Public Works Department. Engineers and other officers with technical duties were vital to the running of the railways. By the late 1930s, more than 40,000 miles of track were being operated in the subcontinent by nearly two thousand officers, around half of whom were Indians and the rest Europeans.[1] This chapter deals with these little-studied cadres in the period 1905–40, focusing on the colonial government's response to Indian politicians' demands for Indianization. It demonstrates that, as in the case of the PWD, colonial officials stressed the necessity of retaining a European kernel of higher engineering and allied posts in the railways. However, in the context of the railways, they justified this more on grounds of security than of gentlemanliness or integrity.

The voluminous academic literature on the railways rarely touches on this issue, although it addresses themes such as the impact of the railways on colonial India's economy, the reception of the railways and their representation in Indian literature, the mobilization of labour in the building of the railways and for their workshops, and the railways'

---

[1] Sources for these numbers appear later in this chapter.

creation of racial and caste-based hierarchies.[2] Like the majority of these studies, the few works that address the history of technical experts in the railways focus mostly on the long nineteenth century.[3] Work on the railways in the twentieth century tends, like other accounts of science, technology, and medicine in India, to follow the arc of the grand narrative of Indian history—entrenched colonialism until World War I, increasingly powerful nationalists and a receding colonial state thereafter. Consequently, they portray the development of the railways in the interwar years as 'dominated by the movements for national freedom and the consequences of [the World Wars]'. Ian Kerr's account of the period 1905–47, titled '"Nationalizing" the Railroads', stresses the growing demands by nationalists to 'make [the railways] more Indian and more responsive to Indian problems, wants and aspirations'. He notes that nationalists demanded more appointments for Indians (Indianization), and the taking over by the state of railway lines run by London-based companies (nationalization).[4] Daniel Headrick draws a direct link between these two demands, writing that 'Indianization began right after [World War I] as the government acquired the various railway companies and influenced their personnel policies'.[5] Yet, as this chapter will show, these valuable

---

[2] See for instance I. D. Derbyshire, 'Economic Change and the Railways in North India, 1860–1914', John Hurd, 'Railways', and Daniel Thorner, 'The Pattern of Railway Development in India', collected in *Railways in Modern India*, edited by Ian J. Kerr (New Delhi: Oxford University Press, 2001); Harriet Bury, 'Novel Spaces, Transitional Moments: Negotiating Text and Territory in Nineteenth-Century Hindi Travel Accounts', Marian Aguiar, 'Railway Space in Partition Literature', and Ian J. Kerr, 'The Railway Workshops and Their Labour: Entering the Black Hole' in *27 Down: New Departures in Indian Railway Studies*, edited by Ian J. Kerr (New Delhi: Orient Longman, 2007); Manu Goswami, *Producing India: From Colonial Economy to National Space* (Chicago and London: University of Chicago Press, 2004), chapter 3.

[3] Ian J. Kerr, *Building the Railways of the Raj: 1850–1900* (Delhi: Oxford University Press, 1995); Ian Derbyshire, 'The Building of India's Railways: The Application of Western Technology in the Colonial Periphery 1850–1920' in *Technology and the Raj: Western Technology and Technical Transfers to India 1700–1947*, edited by Roy MacLeod and Deepak Kumar (New Delhi, Thousand Oaks, and London: Sage Publications, 1995), pp. 177–215.

[4] Ian J. Kerr, *Engines of Change: The Railroads That Made India* (Westport, CT, and London: Praeger, 2007), chapter 6. The quoted text is from pp. 112, 113.

[5] Daniel R. Headrick, *The Tentacles of Progress: Technology Transfer in the Age of Imperialism, 1850–1940* (New York and Oxford: Oxford University Press, 1988), p. 341.

studies neglect the specificities of the Indianization story, which followed different patterns for different departments of the railways, and for Superior as opposed to Subordinate staff.

In addressing these specifics, this chapter aims to enhance understandings of the politics of railway Indianization and the processes by which the state implemented it. While accepting the importance of nationalist politics, it argues also for the significance of incremental changes within the government machinery.

Drawing on government employment lists and commercial directories, the contemporary railways press, engineers' memoirs, official reports and exchanges in the British and Indian Legislatures, I begin this chapter by describing the organizational structure of Indian railways in the period 1905–20. I then chart interwar debates on nationalization and Indianization, showing how they were informed by contemporary views on the importance of engineers on the railways and the abilities of Indians—views which revolved around a conception of the railways as central to the security of the colonial state. Finally, I examine employment statistics to assess the extent to which Indianization took place.

THE RAILWAY BUREAUCRACY, 1905–20

Between 1905 and 1940, India's railway network was among the five largest in the world, along with the United States, Russia, Germany, and Canada.[6] This network was a heterogeneous assortment of railway systems of varying gauges, owned and run by different entities. Of these the most important categories were as follows:

1. Railways owned and managed by the Government of India (referred to simply as 'state railways')
2. Railways owned by the Government of India and managed by a company
3. Railways owned and managed by a company
4. Railways in the princely states (which might be run by the ruler's government or by a company)

[6] Comparison between countries in Table 3.1 on p. 55 of Headrick, *Tentacles of Progress*. The table gives the total length of India's railways in 1940 as 72,144 kilometres, which works out to close to 45,000 miles.

The extent of each type of railway in this period can be gauged from the official statistics for 1920–1. In that year the Government of India operated 8,929 miles of railroad, the Companies 25,211 miles, and the princely states 2,889 miles.[7] Table 4.1 lists the main railways under different categories of ownership and management.

Of the four types of railways, this chapter is especially concerned with (a) state railways and (b) state-owned but company-run railways.

Table 4.1 Ownership and management categories of railways in India, 1920–1

Category	Name of railway
Owned and run by the state	North-Western Railway, East Bengal Railway, Oudh and Rohilkund Railway, Jorhat (Provincial) Railway, Aden Railway
State-owned but company-managed	Bombay Baroda and Central India Railway; Great Indian Peninsula Railway; Madras and Southern Mahratta Railway; Bengal–Nagpur Railway; East Indian Railway; South India Railway; Burma Railways; Assam–Bengal Railway; Rohilkund and Kumaon Railway; Bengal and North-Western Railway (the last two partially company-owned)
Private-owned and company-managed (with domicile of the company in brackets)	[Hyderabad] Nizam's Guaranteed State Railway (England); Martin and Co (India); McLeod & Co (Ind.); Bengal Dooars Railway (England); Darjeeling-Himalayan Railway (Ind.); Barsi Light Railway (Eng.); Assam Railways and Trading Co (Eng.); Guzerat Light Railways (Ind.); Bengal Provincial Railway (Ind.); East India Distilleries and Sugar Factories Ltd (Ind.); Dehri-Rohtas Light Railway (Ind.); Tezpore-Balipara Light Railway (Ind.); Madaya Light Railway (Ind.); Jagadhri Light Ry (Ind.).
Run by the princely states	Railways of Jodhpur, Bikaner, Mysore, Gwalior, Gondal, Bhavnagar, Junagad, Morvi, Udaipur, Navanagar, Dholpur, and Cutch

Source: Railway Board (India), *Railways in India: Administration Report for the Year 1920–21*, vol. I (Simla: Government Central Press, 1922).

[7] Railway Board (India), *Railways in India: Administration Report for the year 1920–21*, vol. I (Simla: Government Central Press, 1922), p. 6.

These categories accounted for the longest and economically the most important lines.

Until 1905, the state lines constituted a Railways branch under the PWD. In that year the government dissolved the PWD's Railways branch, and constituted instead a central Railway Board with regulatory powers over all railways in the country, irrespective of who owned or managed them.[8] Managers of the state railways and the Directors of the company railways would report to the Railway Board. The most senior of the Board's three members was its President. He had equivalent standing to a Government Secretary and reported to the relevant Member (usually the Member for Commerce and Industries) of the Governor-General's Executive Council.[9]

The creation of the Railway Board was the result of a report by Thomas Robertson, who was commissioned by the Secretary of State in 1901 to study the administration of railways in India.[10] This was in response to complaints from the companies that the state—through its consulting engineers, examiners of accounts, and government-appointed directors on the companies' boards—had undue control over 'the management and revenue expenditure' of the company-run railways, whereas they gave the state railways considerable freedom. Robertson stressed the need for uniformity in the running of the railways, whether state- or company-run, and the Railway Board was his scheme for ensuring this.[11]

The result was that from 1905 state and company railways had parallel structures. The 'gazetted' (Superior) officers and 'non-gazetted' (Subordinate) staff of the state-managed railways were mirrored by

[8] Kerr, *Engines of Change*, pp. 25, 75–9; G. Huddleston, *History of the East Indian Railway* (Calcutta: Thacker, Spink and Co, 1906), chapter 1.

[9] *East India (Railway Committee, 1920–21). Report of the Committee Appointed by the Secretary of State for India to Enquire into the Administration and Working of Indian Railways* (Cmd. 1512, London: HMSO, 1921), chapter IV. This report is hereinafter cited as *Acworth Committee Report*.

[10] *Statement Exhibiting the Moral and Material Progress and Condition of India during the Year 1903–04*, p. 130. The full title of Robertson's report is: *East India (Railways). Report on the Administration and Working of Indian Railways. By Thomas Robertson, C.V.O., Special Commissioner for Indian Railways* (Cd. 1713, London: HMSO, 1903).

[11] Robertson, *Report on the Administration and Working of Indian Railways*, chapter 1. The quoted phrase is Robertson's, and appears on p. 13.

the company-run railways' 'covenanted' (contracted) and 'non-covenanted' employees.[12] Both types of railways had similar arrangements in matters of salary and leave.[13]

Railway Departments: Functions and Composition

Most railways, whether run by the state or by a company, were divided into a number of departments as follows:

1. Agency
2. (Civil) Engineering
3. Traffic/Transportation (Commercial)
4. Locomotive and Carriage and Wagon/Transportation (Power)/ Mechanical Engineering
5. Stores
6. Other (including electrical, signal, and bridge engineers)

Each of these departments had a 'Superior Service' and a 'Subordinate Service'. This chapter is concerned with the Superior Services, which comprised the gazetted/covenanted officers (referred to henceforth as 'Superior officers' or simply 'officers'). The following description of recruitment and functions of officers in each department applies mainly to the state railways. Where the company railways differed considerably, attention is drawn to the difference. In general, however, commercial and government lists of employees indicate that the organizational structure of the company lines paralleled that of the

[12] See L. A. Natesan, *State Management & Control of Railways in India: A Study of Railway Finance Rates and Policy during 1920–37* (Calcutta: University of Calcutta, 1946), p. 365n4.

[13] See Natesan, *State Management & Control of Railways*, pp. 368, 386. Although the new system was meant to give companies freedom of operation, the Railway Board retained a degree of control over them. In 1912, for instance, it vetoed the East India Railway Company's proposal to increase its engineers' salaries, partly on the grounds that this might cause dissatisfaction among the employees of other company-managed railways who did not receive a similar raise. 'East India Railway (Pay of Engineers)', House of Commons Debate, 13 November 1912, vol. 43, c. 1970 (Hansard). This and other House of Commons debates cited later were accessed via http://hansard.millbanksystems.com.

state lines.[14] A similar structure was also followed by the railways in the larger princely states.[15]

The first department, the Agency, formed the highest rung of the railway bureaucracy. It comprised the chief managers of the railway (Agents and Deputy Agents on the company railways; General Managers and Assistant General Managers on the state lines). Agents of the company lines reported to the Board of Directors in London (which was further accountable to the Railway Board in India). The General Managers of the state railways reported directly to the Railway Board. Agents and Managers were officers who had risen through the ranks of the other departments, usually Engineering or Traffic.[16]

Civil Engineering, officially listed simply as 'Engineering', was the main executive department. Its officers carried out three broad types of work: maintenance and repair of functioning railroads, known as 'open lines'; survey and construction work on new lines; and, especially for higher-ranked officers, inspection of lines.[17] Within any of these categories, the engineer's workaday duties encompassed a wide variety of activities. Recalling his early days on the railways of Hyderabad state, a British engineer described open-line work as 'largely routine' but nonetheless strenuous. The engineer began his day around seven in the morning and was out on the line until about 6 p.m., when he got back to the office for an hour or so of paperwork. When he was out on inspection, '[a] peon went daily between the engineer and the office carrying a box full of papers for the engineer's orders and signature'. The engineer's usual work consisted of inspecting the track, 'checking stores and labour, measuring work for payment or checking [his subordinate's] measurements'.[18] Construction

---

[14] For example, *India Office List* and *Thacker's Indian Directory*, various years.

[15] See, for example, the list of employees of the Nizam's Guaranteed State Railway, Hyderabad, in *Thacker's Indian Directory* for 1913.

[16] Annexure XIX: 'Railway Department (Revenue Establishments)', *Royal Commission on the Public Services in India: Report of the Commissioners*, vol. I (Cd. 8382, London: HMSO, 1917), pp. 338–44, here p. 388. This document is hereinafter cited as *Islington Commission Report*.

[17] Annexure XVIII: 'Public Works and Railway Department (Engineering Establishment)', *Islington Commission Report*, pp. 326–37, here p. 326.

[18] Omar Khalidi (ed.), *Memoirs of Cyril Jones: People, Society and Railways in Hyderabad* (New Delhi: Manohar Publications, 1991), p. 38.

work, on the other hand, varied according to the specific features of each project. Another engineer, who joined the state railways in 1908, described a project from his early years, when he was posted to the Rajputana country. Charged with bridging a river at a point without a level bedrock, he had to tackle an array of tasks. First, he needed to devise a means of transporting the building material to the site. For this he had to lay a fresh track to the nearest (narrow gauge) station, transport a dismantled broad gauge train from Agra to that point and reassemble it, and then place the broad gauge train on the newly built line leading to the construction site. Meanwhile, he was to design the caissons (airtight chutes leading down to the bottom of the river) and visit Calcutta workshops where they were being produced, most probably by a contractors' firm. 'Then when all had been delivered at the river, I was to get to work training the workmen in the art of working in compressed air, erect the machinery, sink the caissons, and complete the underwater foundations.'[19]

On the state railways, which had earlier been a branch of the PWD, the organization of engineers broadly paralleled that department. Engineer officers were divided into Imperial and Provincial Services. Imperial Service engineers were recruited in two ways. A few were appointed from the Royal Engineer Corps. Others were selected by the Secretary of State from among candidates with an engineering degree from a British college plus a year's practical experience in civil engineering. Candidates with equivalent qualifications were also eligible, provided they had had three years' practical experience.[20] Provincial Service engineers were recruited from the Indian engineering colleges.[21] The term 'Provincial Service' did not imply any relation to the provinces, but denoted only that its members were recruited in India on terms different from those of the Imperial engineers. As in the case of the PWD, this was deemed an unnatural distinction as Provincial and Imperial engineers bore identical responsibilities, and

[19] Victor Bayley, *Nine Fifteen from Victoria* (London: Robert Hale & Company, 1937), chapter V. Quoted text from pp. 88–9. On Bayley's date of joining the railways, see the entry for his name in the Record of Services, *India Office List* for 1922.

[20] Annexure XVIII, *Islington Commission Report*, p. 330. Before 1905, when the railways were part of the PWD, the bulk of these engineers had been recruited from the Royal Indian Engineering College at Cooper's Hill (see Chapter 3).

[21] Annexure XVIII, *Islington Commission Report*, p. 326.

after World War I the two branches were merged into a single Indian Railway Service of Engineers (IRSE), whose officers were recruited in both Britain and India.[22]

On company-run railways, civil engineers were mostly graduates of the British engineering colleges with a few years of practical experience, often on a British railway. For instance, Thomas Guthrie Russell (1887–1963) took an engineering degree in Glasgow University, was apprenticed to a firm of civil engineers in Glasgow for three years, then worked for a further three years on the North British Railway before joining the Great Indian Peninsula Railway as Assistant Engineer in 1913.[23] Cyril Lloyd Jones (1881–1981) graduated from the Central Technical College in London in 1899, then worked for a variety of engineering firms including Mertz and Maclellan, for whom he carried out survey work for a light railway between Maidstone and Chatham, before joining the Nizam's Guaranteed State Railway Co. in Hyderabad in 1904.[24] The company railways usually appointed engineers on renewable three-year contracts. Most engineers did extend their contracts, serving out their entire careers in India (though they might switch railways within the country).[25]

The Traffic Department, also referred to as Transportation (Commercial), was engaged in logistical planning and the control of trains. Officers of this department prepared time tables, dealt with freight, and supervised station masters, signallers, guards, and foremen to ensure the smooth running of trains.[26] Before 1907 Traffic officers on

[22] See *Statement Exhibiting Moral and Material Progress*, 1930–1, pp. 258–60 for steps in railway service reorganization, and Chapter 3 of this book for the analogous reorganization of the public works engineering service. See Annexure XVIII, *Islington Commission Report*, for the rationale behind the amalgamation of the Imperial and Provincial Services.

[23] Obituary, 'Sir Thomas Guthrie Russell', ICE *Proceedings*, vol. 31, no. 1 (May 1965): 126–7.

[24] Khalidi, *Memoirs of Cyril Jones*, pp. 22–8.

[25] Derbyshire, 'The Building of India's Railways', p. 181.

[26] 'Indian State Railways: Regulations for Appointment to the Transportation Department', *India Office List* for 1928, p. 286; Appendix I: 'Courses of Training Prescribed for Probationers in the Transportation and Commercial Departments', *India Office List* for 1928, p. 287. See also 'Indian State Railways: Regulations for Appointment to the Traffic Department' (including Appendix I), *India Office List* for 1910, pp. 228–9.

the state lines were recruited entirely in India; from that year, recruitment for about 60 per cent of the vacancies was done in Britain—a temporary practice begun because of the lack of qualified candidates in India.[27] These officers, who were generally not degree engineers, were selected in England by the Secretary of State from among applicants possessing a university degree, a technical diploma, or at least two years' experience in Traffic Department work on a British or colonial railway along with 'a sound general education'.[28] In addition some officers were appointed from the Royal Engineer Corps. It appears that applicants in India had to be graduates, though it is unclear whether they were required to possess practical experience like their English counterparts.[29]

Mechanical Engineering, or Transportation (Power), consisted of the Locomotive and Carriage and Wagon Departments. Its officers were not degree engineers either. Applicants were required to have knowledge of materials and applied mechanics, and to be able to calculate stresses on and conduct repairs on locomotives and other machinery. They were almost entirely recruited in Britain by the Secretary of State. Locomotive officers were selected from among men with three years' training in railway company workshops and six months' training in running sheds and firing, while Carriage and Wagon candidates were required to have been apprentices in the shops of a railway or builder of wagons, and have had a year's experience as 'outside assistant' on a British line.[30]

Officers of the Stores Department were in charge of procuring, managing, and accounting for supplies. They were recruited entirely in India 'from among candidates "of good education and suitable social position"', but very few Indians were among those selected.[31] Other miscellaneous officers included medical and railway officers

---

[27] Annexure XIX, *Islington Commission Report*, p. 338.

[28] 'Indian State Railways', *India Office List* for 1910, p. 228.

[29] The Islington Commission's report (1917) recommended that candidates in India should be graduates of a university or of Roorkee Engineering College (or possess an equivalent qualification). Annexure XIX, *Islington Commission Report*, p. 339.

[30] 'State Railways: Regulations for Appointment to the Locomotive and Carriage and Wagon Departments', *India Office List* for 1928, pp. 289–90; Annexure XIX, *Islington Commission Report*, p. 339.

[31] Annexure XIX, *Islington Commission Report*, p. 340.

as well as technical specialists (bridge engineers, signal engineers, mining engineers, and colliery engineers). These specialists, on the evidence of the names in government employment lists, were almost all Europeans—probably because training in branches other than civil engineering was not yet being offered at the Indian engineering colleges.[32]

Railway Bureaucracies: Engineering-Centric, Militaristic, and Largely 'European'

Some features of this organizational set-up stand out. The first is that with the partial exception of the Traffic and Stores Departments, the Superior Service was predominantly a cadre of officers with technical qualifications or experience, engaged in technical work. Civil engineers traditionally occupied a pre-eminent position (vis-à-vis the non-engineer officers), their prominence dating back to the pioneer days of the nineteenth century, when the construction of new lines was the most important task.[33] This continued into the early twentieth century: throughout the period under discussion, the Superior officers of the (Civil) Engineering Department were more numerous than those of any other department.[34] In official reports and employment lists, it was usually the Engineering Department that was listed immediately after the Agency, indicating its importance in the organizational hierarchy.

Civil engineers were present on the companies' Boards of Directors, on the Railway Board, and as managers/agents of the state- and company-run railways. For instance, Sir Robert Highet (1859–1934), trained by pupillage in Ayr, joined the East India Railway as Assistant Engineer in 1883, rose to Chief Engineer in 1903, and subsequently Agent in 1912. After returning to England in 1920, he became Chairman of the East India Railway Company and a Director of the Nizam's Guaranteed State Railway Company.[35] Indeed, senior officers from the Engineering Department were sometimes

[32] *India Office List*, various years.

[33] See Kerr, *Building the Railways of the Raj* for the role of civil/construction engineers in the railways in the nineteenth century.

[34] As per statistics presented later in this chapter.

[35] Obituary for Sir Robert Swan Highet, CBE, 1859–1934, in *Minutes of the Proceedings of the Institution of Civil Engineers*, vol. 240 (1935): 787.

placed in charge of other departments like Stores, as in the case of Sir Thomas Guthrie Russell on the Great Indian Peninsula Railway (GIPR) in 1923.[36]

Second, the railways were elaborate, rank-conscious bureaucracies. To get a sense of this, one has only to look at the employment lists of the railways in our period (whether state-run, company-run, or in a princely state), with their myriads of job titles in a variety of grades and classes; strings of titles, professional affiliations, and educational qualifications appended to officers' names; specifics of who was on leave and who was 'acting' in their place; and in some instances salaries.[37] Ian Kerr describes the railway engineers of the latter half of the nineteenth century as 'techno-bureaucrats'.[38] The label holds true for the early twentieth century as well. A. W. C. Addis, the author of a 1910 pocketbook manual for novice railway engineers in India, dwelt at some length upon such functions. When brickfields were set up at a railway construction site, he advised his readers that an assistant should be appointed to control the work and 'keep a record, on a properly drawn up form, of all the bricks carried away, and note their destination'. Likewise, for lime manufacturing depots, the engineer should appoint officials 'whose main duty should be the registering of receipts of every class of material delivered *at* the manufactory, the issue of such material delivered *at* the manufactory and the final issue of the turned out article to works'. If the required forms were not to be had from the department of the Auditor or the Examiner, the engineer must prepare them himself and charge their printing to 'Manufactures'. Bureaucracy in the sense of hierarchy and a proper chain of command was important too. An Assistant Engineer must never delegate the 'fixing of centre-line and cross-line pegs for bridges or buildings' but do it himself; conversely, when he felt that the centre of a bridge as marked on the plan was not ideal, he must never make the change of his own accord: 'Drawings and details must be sent to

[36] He joined the GIPR as Assistant Engineer in 1913, and was made Controller of Stores in 1923. Guthrie Russell went on to become Chief Commissioner, Railway Board, from 1929–40. Obituary, 'Sir Thomas Guthrie Russell', ICE *Proceedings*, vol. 31, no. 1 (May 1965): 126–7.

[37] *India Office Lists* and *Thacker's Indian Directory*.

[38] Kerr, *Engines of Change*, p. 36.

[his superior], [and] in all probability the matter will be settled by his inspection of the site.'[39]

The third significant feature of the railways' Superior Services was the strong military influence on their culture, a feature they shared with public works engineering.[40] However, while in the PWD this was the continuation of a tradition—military engineers had built the earliest public works—military engineers on the railways also performed a strategic function. Historians have noted the strategic importance of railways in frontier regions, and the long-established policy of designing railway stations as fortresses to be protected from potential insurgents.[41] Railway employee lists show that the military presence on the railways (particularly the state railways) in the form of Royal Engineer officers was considerable, especially in the early part of the twentieth century.[42] In 1913, for example, 48 of the 277 state railway officers appearing in the *India Office List* were Royal Engineers.[43] Royal Engineers were present in the Engineering and Traffic Departments, usually having been appointed to the railways soon after receiving their first commission. Such officers spent their entire careers in the railways, although they retained their military rank and were sometimes called away to military service. An example is the career of Captain Cecil Ford Anderson, R. E. (b. 1872), as detailed in a 1910 government dossier. Anderson received his first commission as Second-Lieutenant in 1892 and worked on various state railways from 1895 onwards, occasionally being deputed on military service—as in May–December 1896, when he joined the Soudan Expeditionary Force.[44]

[39] A. W. C. Addis, *Practical Hints to Young Engineers Employed on Indian Railways* (London and New York: n.p., 1910), pp. 59–60 (brickfields), 77 (lime manufacture), 69 (pegs for bridges), 90 (superior's approval).

[40] See Chapter 3 of the present book.

[41] Gyan Prakash, *Another Reason: Science and the Imagination of Modern India* (Princeton: Princeton University Press, 1999), pp. 165–6; Goswami, *Producing India*, pp. 116–17.

[42] Kerr in *Engines of Change* notes the connection between the racial composition of railway staff and security concerns, though he does not mention the Royal Engineers specifically. See p. 118.

[43] *India Office List* for 1913.

[44] Railway Board, *History of Services of the Officers of the Engineer and State Railway Revenue Establishments, Corrected to 1st July 1910* (Calcutta: Superintendent Government Printing, India, 1910), pp. 5–6 [APAC: IOR/V/12/67].

In addition to this, most civil engineers on the railways volunteered for army training. Each railway had its own auxiliary force made up of volunteers to guard its line in the event of conflict. A photograph of (British) officers featuring in a railway magazine, captioned 'E. B. [Eastern Bengal] Railway Battalion Auxiliary Force in Camp at Paksey—24th January, 1930', is typical.[45] In their memoirs, anecdotes abound of the close links between railway engineers and military life. In Hyderabad, Cyril Lloyd Jones got on well with a local army regiment, the 61st Pioneers, even getting them to help occasionally with construction/repair work on the Nizam's Railway lines.[46] Victor Bayley, who joined the railways as an Assistant Engineer in the 1900s, was a member of the Simla Volunteers when working in the Railway Board's office in 1913. When war broke out soon after, he was made a Second-Lieutenant, and then a Temporary Captain on a railway line based in Baghdad. In the 1920s, when he was placed in charge of constructing the Khyber Railway in hostile tribal territory in north-western India, Bayley's base was inside a military camp.[47] The maintaining of a military character in the railway bureaucracies was a conscious strategy and assumed particular significance in decisions on Indianization, as we shall we see later in this chapter.

The fourth, and politically the most important, feature of the railways' Superior Services was their racial composition. The Superior officer positions, both in state railways and company-run railways, were dominated by Europeans. Anglo-Indians, despite their large presence in the railways as a whole (around 11,000 employees in the 1920s, according to one estimate),[48] were a minor presence in the Superior Services. In 1913, of 447 officers in the state railways drawing a salary of Rs 200 per month or more, only forty-five were 'native' Indians and seventy-two Anglo-Indians, a combined share of 26.17 per cent (see Table 4.2). The higher the salary bracket, the more heavily European-dominated were the posts.

[45] *Indian Railway Gazette*, March 1930, p. 79.

[46] Khalidi, *Memoirs of Cyril Jones*, p. 52.

[47] Bayley, *Nine Fifteen from Victoria*, pp. 222–4, 236, 282.

[48] Frank Anthony, *Britain's Betrayal in India: The Story of the Anglo-Indian Community* (Bombay: Allied Publishers, 1969), p. 95.

Table 4.2 Salaries of Superior Officers, State Railways, 1913

Salary per month	Europeans	Indians		Total	Total officers	Anglo-Indians (%)	'Native' Indians (%)	'Natives' + Anglo-Indians (%)
		Anglo-Indians	Other ('native') Indians	Total				
≥Rs 200	330	72	45	117	447	16.11	10.07	26.17
≥Rs 500	257	42	19	61	318	13.21	5.97	19.18
≥Rs 800	155	17	11	28	183	9.29	6.01	15.30

Source: Based on tables in *Royal Commission on the Public Services in India. Report of the Commissioners*, vol. I (Cd. 8382, London: HMSO, 1917) [*Islington Commission Report*], pp. 24–5.

Note: The category of 'native' Indians here includes Burmans, while the absence of a category for Domiciled Europeans in the source suggests that they are counted along with the Anglo-Indians.

Not only did Europeans form an overwhelming majority in the higher officer positions, they were also—as Chapter 3 showed for the PWD—paid higher salaries and offered greater benefits than their Indian counterparts. In the 1910s, the scale for the Imperial Service (mostly Europeans) in the ranks of Assistant Engineer and Executive Engineer began at Rs 380 a month and went up to Rs 1,250. The corresponding figures for the Provincial Service (recruited in India) were Rs 250 and Rs 900.[49] Even after the Imperial/Provincial distinction was eliminated and new pay scales put in place (1920–1),[50] engineers of 'non-Asiatic domicile' were paid a substantial overseas allowance *beginning* at Rs 150 per month—as was the case in other All-India and Central Services. Other privileges applied across the services, including a number of paid passages to England, and arrangements for European officers and their families to be attended by European doctors.[51] Similar disparities are likely to have been present on the company-run lines, whose personnel policies, as we have seen, closely mirrored those of the state railways.

Indeed, pay was just one element in a deeply hierarchical culture, which extended to housing and leisure. This was evident in the geographical layout of railway settlements, with their rank-wise neighbourhoods, amounting to separate European and Indian sections.[52] A new arrival to the eastern railway town of Burnpur in 1921 described

[49] Annexure XVIII, *Islington Commission Report*, esp. p. 333.

[50] Railway Board, *Railways in India: Administration Report for the Year 1920–21*, vol. 1, p. 31.

[51] *East India (Civil Services in India). Report of the Royal Commission on the Superior Civil Services in India* (Cmd. 2128, London: HMSO, 1924), pp. 29–31, 35–6, 53. This source is hereinafter cited as *Lee Commission Report*.

[52] Indian Superior Service officers may have lived alongside the Europeans. This is my inference from conversations with descendants of Indian railway officers, and from the following description of a railway town in the 1930s by an Indian engineer: 'Jamalpur was a Railway colony housing British officers and Anglo-Indian supervisors with a sprinkling of Indian officers and supervisors' (D. V. Reddy, *Inside Story of the Indian Railways: Startling Revelations of a Retired Executive* [Madras: M. Seshachalam, 1975], p. 17). Headrick in *Tentacles of Progress* (p. 342) also quotes this line. See also Laura Gbah Bear, 'Miscegenations of Modernity: Constructing European Respectability and Race in the Indian Railway Colony, 1857–1931', *Women's History Review*, vol. 3, no. 4 (1994): 531–48. Bear writes that '[r]ailway company officials' had a 'vision of railway colonies as artificial European enclaves designed to protect their residents, domiciled Europeans and Eurasians, from the temptations of India' (p. 531).

its European section as 'a perfect little model village' in a six-by-six grid of avenues and streets. The railway yard and a golf course occupied two corners of this grid. The bungalows were spacious, and apart from the servants who lived at the back of these, 'no other natives live[d] on the Station [European section]'. Within the Station, the Superior officers were segregated from the non-officer employees recruited from England for the workshops. The Indian workmen were on the other side of the railway line, in 'a typical native bazaar'. The Station contained a 'Club, of which all members of the superior staff and their families [were] ex-officio members and no one else'.[53] Railway towns usually had separate clubs and Railway Institutes (recreation centres for employees) for Europeans and Anglo-Indians on the one hand, and Indians on the other.[54] These differences continued until the eve of Independence.[55]

From the 1910s, nationalist politicians demanded two types of reform in the railways: first, in the way they were managed, and second, in the composition of their officers. These form the focus of the next two sections.

NATIONALIZATION: DEMANDS AND RESPONSE

The seeds of debate about the ownership and management of India's railways had been sown in the late nineteenth century, when the original contracts of some of the railway companies with the Government of India had run their course. The Government bought these companies' assets, but invited the companies to continue managing the railroads in question.[56] In the early decades of the twentieth century, the anomalies created by this system began to provoke protests.

[53] Philip Napier, *Raj in Sunset* (Ilfracombe, Devon: A. H. Stockwell, 1960), pp. 29–30.

[54] This was still the case on the Bengal and Assam Railway in 1942. See Question 175, *The Legislative Assembly Debates (Official Report)*, vol. III, 1942 (Delhi: Manager of Publications, 1942), p. 46 [APAC: Microfilm IOR Neg 14795]. On Railway Institutes, see Anthony, *Britain's Betrayal in India*, p. 357.

[55] Jogendra Nath Sahni, *Indian Railways: One Hundred Years, 1853 to 1953* (New Delhi: Ministry of Railways [Railway Board], 1953), p. 125.

[56] See Kerr, *Engines of Change*, pp. 75, 77.

Some of these anomalies were related to macroeconomic policy. In 1916 Sir Guilford Molesworth, former Government Consulting Engineer for state railways, argued that having a mix of state- and company-managed railways was inimical to the country's economic interests. The railway companies, he said, operated solely to maximize the profit of their shareholders and not to aid economic activity in India. They set very high rates for freight traffic, and where goods had to pass through the territories covered by their lines, Indian export trade was adversely affected.[57]

But the main protests against company management came from Indian thinkers who did not feel that the railways were adequately representative of the interests of the Indian public.[58] A pamphlet circulated by a solicitor of the Bombay High Court, Faredun K. Dadachanji, in 1920 discussed the various reasons why Indians opposed the system of company management. They felt that the Railway Board was partial to the interests of the companies; that Railway Board officials went on to become directors on the boards of the railway companies upon retirement, causing a potential conflict of interest; that Third Class passengers and those on pilgrimage specials were ill-treated and transported in the worst possible conditions; and that Indians were kept out of the well-paying jobs on the company railways. In sum, 'Company management of the railways already owned by the nation' was '[a] most monstrous and a most insane doctrine'.[59]

Discontent about the administration of the railways (especially the company-run lines) peaked in the aftermath of World War I, the stresses of which placed the railways in an economically precarious condition. Trains had to take on unprecedented freight loads, locomotives were despatched for use in theatres of war in Iraq, and (as discussed in the next chapter) the Indian-owned Tata Iron and Steel

[57] Guilford L. Molesworth, *Indian Railway Policy*, Indian Railway Series no. 1, compiled and edited by Faredun K. Dadachanji (Bombay: F. K. Dadachanji, 1920), esp. pp. 1–2.

[58] Kerr, *Engines of Change*, p. 113.

[59] Dadachanji, 'Compiler's Introduction' in Molesworth, *Indian Railway Policy*, pp. ix–xlii. The quoted text is from p. xlii. As the other two categories of railways—lines owned *and* run by companies, and lines in the princely states—accounted mostly for the smaller, economically less important networks, the debate was not particularly concerned with them.

Company stepped in to supply steel rails to the government at special prices. By 1920 'Indian opinion on the question of state management of railways was unanimous and the demand for state management was reinforced both on political and economic grounds'.[60] Political, because state-run railways would be more 'Indian' than British-company-run ones; and economic, because initiatives on the scale required for the post-war repair of the railways' fortunes could only come from the state.

In November 1920 a committee was set up under the chairmanship of Sir William Acworth (a lawyer who had sat on the 1916 Royal Commission on Canadian Railways) to look into the issue. Foremost among the terms of reference of the Acworth Committee was to recommend what system of management would work best for the company-run (but state-owned) railways when their contracts with the government came to an end. The most immediate case was the East Indian Railway Company, whose contract had been due to end in 1919 and was subsequently extended to 1924.[61] The Committee was to assess 'the relative advantages, financial and administrative,' of management by the state and management by a company (domiciled in England, India, or a combination of the two). The ten-member Committee (comprising seven British and three Indian members) was also charged with suggesting ways to streamline the structure and working of the Railway Board and methods of raising the funds necessary for the running of the railways in India.[62] According to the evidence placed before the Committee, there was a chronic shortage of funds available for maintenance work, let alone fresh construction. This was primarily a result of the fact that the railways did not have

[60] Headrick, *Tentacles of Progress*, p. 85; Kerr, *Engines of Change*, p. 115 and p. 120; *Acworth Committee Report*, p. 19; Natesan, *State Management & Control of Railways*, pp. 11–12, 12n15. The quote is from Natesan, p. 12.

[61] *Acworth Committee Report*, p. 3. Most railway companies had begun in the mid-nineteenth century as guaranteed companies (the colonial government guaranteed them a minimum of 5 per cent return on investment per annum) on contracts lasting fifty or sixty years. The Government had a right to buy the assets of these companies after about twenty-five years, which it did, though in most cases it asked the concerned company to continue managing the railway (*Acworth Committee Report*, pp. 60–1, paragraphs 187–90).

[62] *Acworth Committee Report:* 'Terms of Reference' [front matter] and p. 3.

an independent budget, and a reorganized Railway Board would enable the separation of railway finances from the government's general budget.[63]

At the end of their enquiry, the members were unanimous in concluding that the management of Indian railways should not be carried out from London. The directors of the railway companies in London might have 'technical and expert knowledge', but they were far removed from the actual field of operation, and would not find it easy to keep abreast of 'the modern social and trade conditions of India'.[64] But the Committee was divided upon the issue of who should actually be allowed to manage the railways. Five members, including the Chairman (generally referred to as the majority, as the Chairman had the casting vote), recommended that the state take over and manage the company-run lines when the respective contracts expired. The remaining five were in favour of setting up railway companies with Indian domicile to run the state-owned railways.

The majority included the Chairman (William Acworth), V. S. Srinivasa Sastri (Member of the Viceroy's Council of State), E. H. Hiley (formerly of the British and later New Zealand Government Railways), Purshotamdas Thakurdas (representing Indian business interests), and James Tuke (Director of Barclays Bank and of the British Linen Bank). They began by arguing that under the existing system companies had no real autonomy, for they managed railroads they did not own, and were subject to government control via the Railway Board. They did not feel that railway officers could function efficiently 'with a divided allegiance to the board of directors which appoints and pays them, and to the Government authority which stands behind the directors'. On the other hand, they had found that the managements of state railways were as dynamic and open to 'new methods' as company-run railways. On another note, the members in the majority laid great stress on the importance of respecting Indian public opinion, which was 'practically unanimous' in favour of management directly by the state. They pointed out that Indian taxpayers had funded the building of the railways; that the Indian public were their main customers; and that the Legislative Assembly in Delhi was the body

[63] *Acworth Committee Report*, pp. 20, 36.
[64] *Acworth Committee Report*, p. 64.

that approved the railway budget. In any case, they did not want to antagonize the masses: even if state management did not result in an improvement, they argued, 'we are quite sure that its failures would be judged more leniently by the Indian public'.[65] The railways, in this view, were more a public good than a set of commercial enterprises.

The minority was made up of Sir Henry P. Burt (Chairman of two railway companies and a former President of the Railway Board), Sir Rajendra Nath Mookerjee (Calcutta-based engineer-businessman), Sir Arthur R. Anderson (Railway Board President, 1919–20), Sir George C. Godfrey (Agent of the Bengal–Nagpur Railway Company), and Sir Henry Ledgard (representing European business interests in India). They saw democracy as a threat to the business-like running of the railways. The railways, they felt, were 'primarily commercial under-takings' that should be run by companies 'so as to secure economy and efficiency'. Burt and his colleagues felt that the new political situation after the Montagu–Chelmsford Reforms of 1919 would accentuate the risks inherent in state-run railways. As the process of democratiza-tion advanced, India '[would] be faced with the dangers which lurk in the handling of a commercial business through the agency of a State service and which experience elsewhere condemns'. In particular, the Committee's minority were concerned that state management would lead to inflexibility: for instance, promotions might be made based solely on seniority without considering candidates' 'efficiency or local knowledge'.[66]

The trade journals that catered to senior European railway offic-ers reflected these officers' view of the railways as businesses, and their abhorrence of the intrusion (as they saw it) of political considera-tions into railway policy and administration. The British-based *Rail-way Gazette*, reviewing the Acworth Committee Report, poured scorn on the members supporting nationalization, asking if it would at all be possible to 'infuse and maintain in a Government department a spirit which Government departments rarely possess—a commercial spirit...'[67] Another journal, the *Indian Railway Gazette*, lamented the

---

[65] *Acworth Committee Report*, pp. 65–73.

[66] *Acworth Committee Report*, pp. 73–5.

[67] *The Railway Gazette*, 21 October 1921, pp. 611–12.

fact that the railway companies had not done enough to press their case against nationalization. 'The whole question, by a skilful handling of the Indian Press, has been turned into a political affair, and once on that field of debate the matter became more than a one-sided question, seeing that the preponderating influences in the Assembly are Indian.'[68] In other words, political considerations were external and irrelevant influences that would cloud the issue. The question should be decided on the basis of business principles: what would make the railways more profitable?[69] The reference to 'preponderating [Indian] influences' also betrayed a fear that qualified European men of business and technical experience would be outvoted by uninformed Indian politicians.

In this climate of establishment opinion, William Acworth felt the need to clarify his reasons for supporting nationalization. In a letter to the *Times*, he insisted that while his Indian colleagues Srinivasa Sastri and Purshotamdas Thakurdas 'were doubtless much influenced by the almost unanimous Indian opinion in favour of direct State management, ... the decision of my two English colleagues and myself was mainly based on different grounds'. Acworth wrote that he had merely chosen the lesser evil in the circumstances: 'The actual choice being between companies with no freedom of initiative, bound to refer and defer on every occasion to their owner and master the State, and direct State management, I unhesitatingly voted for the latter.'[70]

In any case the die was cast; the Legislative Assembly in Delhi endorsed the majority recommendations of the Acworth Committee, and nationalization began. In 1925 the state took over two of the largest company-run lines, the East Indian Railway (EIR, based in Calcutta) and the Great Indian Peninsula Railway (GIPR, based in Bombay).[71] The government repeated the process when each company's

---

[68] *Indian Railway Gazette*, April 1923, p. 108.

[69] A similar point was made by Sir Thomas Catto, speaking at the half-yearly general meeting of the Bengal Coal Co. Ltd, Calcutta, on 21 December [1922]. The address was reprinted in the *Indian Railway Gazette*, February 1923, pp. 47–8.

[70] Quoted in *Indian Railway Gazette*, April 1923, p. 124.

[71] *Report by the Railway Board on Indian Railways 1924–25*, vol. I (Calcutta: Government of India Central Publication Branch, 1925), pp. 4–5.

contract came to an end; by 1944 nearly all the railways in India had been nationalized.[72]

INDIANIZATION: DEMANDS AND RESPONSE

Indian politicians saw nationalization as a way to hasten the Indianization of the railways, but the demand for Indianization had its own wider context and history. As early as 1878–79, a Select Committee of the British Parliament had suggested employing fewer Europeans as a way to make the maintenance of state railways more economical, but the railways were reluctant to do this.[73] In the twentieth century, political arguments were added to the economic ones. Indian politicians' demands for greater training and employment opportunities for Indians resulted in several government enquiries into the Indianization question.[74]

The Islington Commission's 1917 report suggested varying measures to increase opportunities for Indians in different departments of the railways. For Engineering, the Commission recommended a system of dual recruitment in England and India in equal proportion. In the Traffic Department Europeans might be selected from among Royal Engineer officers, and for the rest preference should be given to qualified candidates from India. In the almost entirely European departments of Locomotive and Carriage and Wagon, the railways should groom (Indian) subordinates for promotion, and invest in developing institutions to provide training, a potential locus for which already existed in the country's many railway workshops.[75]

The Acworth Committee of 1920–1 also touched upon Indianization in its report on the question of nationalization of the railways. Its members continued the Islington Commission's emphasis on the need for new training facilities for the Traffic and Mechanical

[72] Kerr, *Engines of Change*, p. 121.

[73] Headrick, *Tentacles of Progress*, p. 322.

[74] See Aparna Basu, 'Technical Education in India, 1854–1921', in *Essays in the History of Indian Education* (New Delhi: Concept, 1982), pp. 39–59, here p. 52; John F. Riddick, *The History of British India: A Chronology* (Westport, Connecticut: Praeger, 2006), pp. 94, 100–1.

[75] *Islington Commission Report*, pp. 22–3; Annexure XVIII, *Islington Commission Report*, pp. 328–9; Annexure XIX, *Islington Commission Report*, pp. 338–9.

Engineering (Locomotive, Carriage and Wagon) Departments. They suggested that the government set a target of a certain percentage of Indians within a specified time frame (though they did not themselves specify any numbers).[76]

The Lee Commission of 1923–4, whose task was 'to accelerate Indianization' while simultaneously maintaining a European component to prevent any abrupt change in the character of the services,[77] recommended that recruitment for the Superior Services of the state railways continue to be conducted both in England (by the Secretary of State) and in India (by the central government).[78] Their specific recommendations on Indianization targets were as follows:

> *State Railway Engineers—Superior Revenue Establishment, State Railways—* We understand from the evidence placed before us that the present rate of recruitment (taking an average over the departments as a whole) has been designed with a view to securing, *as soon as practicable*, a cadre of which, out of every 100 officers, 50 shall have been recruited in India and 50 in Europe. The date at which this cadre may be reached is, we are informed, dependent on the provision of adequate training facilities in India. Measures with that end in view were advocated by the Islington Commission and we are informed that facilities have already been provided to a limited extent. We are strongly of opinion that the extension of the existing facilities should be pressed forward as expeditiously as possible in order that recruitment in India may be advanced *as soon as practicable* up to 75 per cent of the total number of vacancies in the railway departments as a whole, the remaining 25 per cent being recruited in England.[79]

Thus, for every officer recruited in Britain, three must be recruited in India. This applied to the Superior Services of the state railways 'as a whole', that is, across all its departments. This ratio, and the larger aim of an overall staff proportion of 50–50, were to be arrived at 'as soon as practicable' rather than within a set time frame.

---

[76] *Acworth Committee Report*, p. 58.

[77] *Lee Commission Report* , pp. 1 and 4–6. The quoted text is from p. 6. Indianization was already taking place, but '[i]n Indian political circles ... the rate of Indianisation adopted since 1919 was regarded as illiberal' (*Lee Commission Report*, p. 6).

[78] *Lee Commission Report*, p. 10.

[79] *Lee Commission Report*, p. 23 (paragraph 42d). Emphasis mine.

One point is crucial in the translation of this recommendation into policy: the figure of 75 per cent was taken to refer to *Indians*, rather than officers recruited *in India*.[80] In fact, the two categories were virtually identical, as the Government of India determined in 1926 that henceforth only statutory Indians would be recruited in India and Europeans in Britain (with two exceptions: in the Engineering Department this would be implemented from 1927, and in Mechanical Engineering and Transportation [Power], from 1933).[81] As we shall see in the next section, government reports in the following years gave statistics showing employees categorized as European or (statutory) Indian, rather than on the basis of where they had been recruited.

Significantly, the 75 per cent formula was also accepted by the company-run railways.[82] In the following decades the Lee prescription of 3:1 recruitment was used as the definitive benchmark against which to measure Indianization.

Mutual Mistrust on Indianization

Although Indianization targets for the railways had received the government's seal of approval, European sections of the railway community were apprehensive, questioning Indians' readiness for responsible positions in terms of their competence and loyalty. This reaction was similar to the PWD case, although the reasons—implicit

[80] In a speech in 1924 the Viceroy said, referring to a recent debate on railway finance in the Legislative Assembly, that '[t]he Lee Commission had made recommendations on this question [Indianization] ... and before the debate on railway finance in the Assembly, the Government of India had decided to accept these recommendations which have the effect of pressing forward as rapidly as possible the extension of existing facilities in order that "the recruitment *of Indians* be advanced as soon as practicable up to 75 per cent of the total number of vacancies in the railway department as a whole"' (Appendix F: 'His Excellency the Viceroy's address to the Indian Railway Conference Association at the Opening of the 1924 session', in *Report by the Railway Board on Indian Railways for 1924–5*, vol. I, pp. 101–5, here p. 104). Emphasis mine.

[81] Reported in *Indian Engineering*, 24 July 1926, p. 49.

[82] C. A. Innes to A. K. Acharya, in *Extracts from the Debates in the Indian Legislature on Railway Matters, Delhi Session—January to March 1927* (Calcutta: Government of India Central Publication Branch, 1927), p. 118 [APAC: IOR/V/25/720/45]. Natesan's 1946 study adds that the Lee Commission's 'Concessions' (that is, recommendations on pay and leave) 'were faithfully copied on company-managed railways'. Natesan, *State Management & Control of Railways*, p. 386.

and explicit—given for it in the railways were slightly different. Chapter 3 showed that Indians' gentlemanliness, courage, and other non-technical attributes were questioned. In the railways, two main parameters were used in the debate on employing more Indians: first, their 'efficiency', which was never clearly defined, but appeared to refer to a mix of technical and non-technical attributes; and second, their 'loyalty', that is, whether or not their political sympathies lay with the colonial state.

The concern with 'efficiency' was a long-standing one. That Indians lacked the aptitude and temperament required for railway work was a stereotype current in the nineteenth century;[83] it was still present in the twentieth. It was clearly visible in the railway press, a vivid example being a series of cartoons that originally appeared in the Great Indian Peninsula Railway Magazine and was published in book form, c. 1921, under the title *The Koochpurwanaypore Swadeshi Railway* (which might be translated as the Anything-Goes-Town Indigenous Railway).[84] The satirist, working under the pseudonym of 'Jo Hookm' ('as you command'), presented a wry vision of what might happen if the railways were to be completely indigenized. In one of the drawings, an Indian Chief Engineer, 'Mr. Sleeper Rastaji', is shown in his office.[85] Measuring instruments and an incomplete blueprint languish in the background, while the dissolute engineer lounges on the floor, hookah in hand, entertained by minstrels and dancing girls. The cartoonist saw the Indian engineer as not only weak of character, but also as incapable of grasping the basic principles of construction: another drawing shows workmen erecting a bridge that the fictional 'K.P.R. [Koochpurwanaypore Railway] Bridge Engineer has designed ... on the quite novel principles, being a combination of the Cantilever, Truss and Suspension all rolled into one (Patent applied for)'.[86]

---

[83] See Manu Goswami, *Producing India*, chapter 3; Headrick, *Tentacles of Progress*, chapter 9; and Derbyshire, 'The Building of India's Railways', p. 184.

[84] 'Jo Hookm', *The Koochpurwanaypore Swadeshi Railway*, 2nd edn (Calcutta and Simla: Thacker, Spink & Co.; Bombay: Thacker & Co. Ltd, c. 1921). I am indebted to the gloss on the following webpage of the Indian Railways Fan Club for drawing attention to the fact that 'Koochpurwanaypore' is an Anglicization of 'kuch-parvah-nahi-pur'. Indian Railways Fan Club, 'Books, Timetables, Videos etc.—II', available at http://www.irfca.org/faq/faq-books2.html (accessed on 3 September 2015).

[85] 'Jo Hookm', *The Koochpurwanaypore Swadeshi Railway*, p. 22.

[86] 'Jo Hookm', *The Koochpurwanaypore Swadeshi Railway*, p. 29.

A similarly disdainful pronouncement came from a visiting American engineer who wrote to the *Indian Railway Gazette* (a magazine whose articles were usually aimed at the European section of the railway community) in 1923. He was impressed by the young (European) railway employees he had come across, but felt they were under threat of being swamped by incompetent Indians if they tried to 'cut in by too much of this Indianisation'. Indians were 'peculiarly lacking' in the 'instinct' required for running the railways (the writer did not specify what this instinct was).[87] The *Gazette*'s own view of Indians was in subtle agreement with this. Commenting in 1930 on debates on the railways in the Indian Legislative Assembly, the magazine spoke of the need for the Railway Board to keep the 'balance' between efficiency and Indianization—implying that the one could only be achieved at the expense of the other. This balance, it opined, had thus far been 'held with splendid British impartiality and fair play'.[88] Six years later, the *Gazette*'s tone had grown more strident. Launching a tirade against the Railway Board's purported inability to deal with competition (particularly from roadways), it thundered that the Board

> will do nothing but sit on their hunkers and waste time at conferences knowing full well that nothing of practical value will result. If they had even a pinchbeck Napoleon on the Board he might try and do something, but the existing body has the mentality of ordinary N.C.O.'s, when unfortunately they should have the vision and daring of an Allenby.

Far from being decisive, though, 'this Railway Board has become a body more concerned with keeping itself in the good graces of the Legislatures than improving and maintaining India's Railways and hence much of its time is spent in issuing inspired statements as to the wonderful work it has done, when in truth it is a most ineffective body'.[89] Governance through legislatures (which, needless to say, were concerned with matters like Indianization) was thus cast as the antithesis of military efficiency.

Prejudice aside, the equation of Indianization with inefficiency may partly be ascribed to senior engineers' and managers' lack of

[87] *Indian Railway Gazette*, December 1923, pp. 445–6.

[88] *Indian Railway Gazette*, April 1930, p. 123.

[89] *Indian Railway Gazette*, July 1936, p. 142.

confidence in the facilities available to Indians for technical training in railway work. For railway engineers in particular, having some
practical experience prior to receiving an appointment was seen as
essential. As we saw earlier in this chapter, most British engineering recruits had already had some experience on British railway lines.
On the other hand Indian graduates, whether of Indian or British
engineering colleges, '[had] difficulty in obtaining that initial practical
experience which is an essential complement to their college course
before they can be considered competent to practise as engineers'.[90]
That this applied also to Indians in Britain was not unusual: the
Indian Industrial Commission (1916–18) had observed that Indian
students seeking practical experience in Britain were usually turned
down.[91]

The second argument against Indians was closely related to the
military character of the railways. This was that Indians' loyalty to
the colonial regime was suspect. The equation of the railways with
the security of Britain's Indian empire, both internal and external,
was an acknowledged article of policy. The Secretary to the Railway
Board had told the Islington Commission that the railways were 'a
necessary factor in maintaining the security of the country'.[92] The
conspicuous presence of Royal Engineer officers in the Engineering
and Traffic Departments was a way of ensuring that the strategically
important railways remained in the control of combat-ready officers loyal to the colonial government. Official reports said as much,
directly or indirectly. In recruiting for the Traffic Department, for
example, it was essential 'owing to considerations of policy', 'to maintain a nucleus of officers imported from Europe', a solution to which
was 'the appointment in India of Royal Engineer officers'.[93] In the

[90] *Statement of Moral and Material Progress of India*, 1926–7, pp. 173–4.

[91] Indian Industrial Commission (IIC), 1916–18, *Report* (Calcutta: Superintendent
Government Printing, 1918), p. 107. According to the IIC, an earlier enquiry had
suggested that the industries in which Indians sought apprenticeships were wary
of '[assisting] possible competitors' (Indian Industrial Commission [IIC], 1916–18,
Report, p. 107).

[92] Quoted in Hirday Nath Kunzru, *The Public Services in India (Political Pamphlets—
II)* (Allahabad: Servants of India Society, 1917), p. 149.

[93] Annexure XIX, *Islington Commission Report*, p. 338.

1920s, Indianization notwithstanding, recruitment in England by the Secretary of State could not be terminated for 'those *quasi* security Services, the Railways, the Telegraph Engineers, the Customs, and Political ... [which] are the Central Services upon which the military security and commercial credit of the country depend'.[94]

Among the chief concerns of British parliamentarians debating Britain's policy in India was the maintaining of a minimum quota of Anglo-Indian employees (in all grades) on the Indian railways, which would allow the raising of auxiliary forces loyal to the colonial government in the event of rebellion.[95] Indianization, in their view, would reduce the government's sources of military manpower and make the railways, and in consequence the empire, susceptible to sabotage.[96] Speaking in the House of Commons in 1933, the Duchess of Atholl drew attention to 'the need for securing a disciplined British reserve for the management of the railways, in case there should at any time be a failure of the Indian staff to run the railways efficiently and loyally'. This reserve staff must contain Britons capable of fulfilling the roles of subordinates as well as officers. Following the Punjab unrest of 1919, the Duchess said, the Government of India began to '[train] every year some officers and men of the Royal Engineers in the management of locomotives on the Indian railways'. But that practice had not been continued in recent years. The Duchess asked if 'with a decreasing number of Europeans and Anglo-Indians on the staff of the railway, and without any of the trained British reserves to man the locomotives, the defences of that great country [could] be adequately secured'.[97]

[94] 'Minute by Sir Reginald Craddock upon certain of the conclusions of the commission', *Lee Commission Report*, pp. 127–38, here p. 135.

[95] As shown by some of the arguments in 'Clause 231. (Application of preceding section to railway services, and officials of courts.)', House of Commons Debate, 4 April 1935, vol. 300, cc. 547–81 (Hansard).

[96] See also David Campion's detailed account of the role of the railway police, in which he mentions the use of auxiliary forces made up of Europeans and Anglo-Indians working in the railways. David A. Campion, 'Railway Policing and Security in Colonial India, c. 1860–1930', in *Our Indian Railway: Themes in India's Railway History*, edited by Roopa Srinivasan, Manish Tiwari, and Sandeep Silas (New Delhi: Foundation, 2006), pp. 121–53, esp. p. 147.

[97] 'India Office', House of Commons Debate, 17 July 1933, vol. 280, cc. 1549–608, here cc. 1607–8 (Hansard). While the Duchess's intervention is quoted here to show

In the light of such views, it is not surprising that nationalists in the Legislative Assembly in Delhi were unconvinced by the government's undertakings on Indianization. They questioned the representatives of the Railway Board minutely, especially on whether they were meeting the Lee Commission's target (75 per cent Indian recruitment for the Superior Services). The Board repeatedly assured members of the Legislature that the Lee Commission's recommendations were being implemented. Sir Charles Innes, Railway Member of the Governor-General's Council, told legislator A. K. Acharya in 1927 that regulations had been published for the recruitment in India of Superior Service officers in the state railways' Engineering and Transportation Departments, and that competitive examinations had been held in November of the previous year. Regulations for Signal and Bridge Engineering were 'under consideration' (as we observed earlier, these positions were virtually all held by Europeans).[98]

Distinctions in pay between British and Indian engineers were also questioned. Lt-Col H. A. J. Gidney, a prominent representative of the Anglo-Indian community,[99] asked A. A. L. Parsons (Financial Commissioner of the Railway Board) in 1930 if there was no fixed starting salary for officers recruited as signal engineers in England, and whether all such officers were recruited on higher starting salaries than those selected in India. Parsons replied that 'there is a fixed incremental scale of pay for Engineers but Government reserve the right to appoint officers whether recruited in England or in India,

---

the establishment's attitude towards Indian employees, her fears were ill-founded in statistical terms. Although she was right about the fall in European numbers overall, the combined share of Europeans/Domiciled Europeans/Anglo-Indians in the Locomotive and Carriage Departments as of 1933 was still extraordinarily high: 84.35 per cent of officers and 82.28 per cent of Upper Subordinates. Appendix F, *Report by the Railway Board on Indian Railways for 1933–34*, vol. I (Delhi: Manager of Publications, 1934) and Appendix G, *Report by the Railway Board on Indian Railways for 1924–25*, vol. I.

[98] C. A. Innes's reply to B. Das, in *Extracts from the Debates in the Indian Legislature on Railway Matters: Delhi Session—January to March 1926* (Calcutta: Government of India Central Publication Branch, 1926), p. 80 [APAC: IOR/V/25/720/44]; C. A. Innes to A. K. Acharya, in *Extracts from the Debates in the Indian Legislature on Railway Matters, Delhi Session—January to March 1927*, p. 118.

[99] On Gidney's role in representing Anglo-Indians, see Anthony, *Britain's Betrayal in India*, pp. 87ff.

having regard to their age and qualifications, on initial rates of pay higher than the minimum of the prescribed scale'. To the second part of the question, the reply was 'in the negative'.[100] On other occasions the scrutiny resulted in tense exchanges. When B. Das asked in 1926 if Government would implement 'their promise to this House of 75 per cent Indianization' in selection—specifically of engineers—in England and in India, Innes replied: 'I should not have thought it necessary for the Honourable Member to ask that question, because when Government give a promise, they invariably carry it out.'[101]

Through the 1920s and 1930s the government certainly made efforts to show that they were attempting to keep their 'promise', furnishing statistics on Indianization in annual reports. But did they actually keep it, or was the Indian politicians' scepticism justified? What was the result of the tug of war between political pressure in the Legislative Assembly and opinion in sections of the railway bureaucracy? To answer these questions it is necessary to examine the numerical progress of Indianization of the railways' Superior Service, the measures taken by the government to provide new facilities for training, and the priorities it revealed in the process.

THE EXTENT OF INDIANIZATION, 1920–40

In assessing the extent of Indianization in the 1920s and 1930s, I use statistics compiled from two main sources: the annual administration reports of the Railway Board and the *Statement Exhibiting the Moral and Material Progress and Condition of India*, tabled each year in the British Parliament by the India Office. I also base qualitative observations on a perusal of government and commercial employment lists.[102] An assessment of the training facilities introduced in this period (especially for Indians) is made on the basis of the qualitative sections of the Railway Board's annual Administration Reports.

[100] Printed in the *Indian Railway Gazette*, May 1930, p. 194.

[101] *Extracts from the Debates in the Indian Legislature on Railway Matters: Delhi Session—January to March 1926*, p. 20.

[102] *India Office List* and *Thacker's Indian Directory*. The former of these sources refers to an annual publication of the India Office in London, listing the officers appointed to the various all-India government services. The latter was a commercial almanac produced annually by Thacker, Spink & Co., a Calcutta publishing firm. Continuous series of both publications are available on the open shelves of the APAC.

In understanding the progress on Indianization in statistical terms, two different measures are important. One is the *composition* of the Superior Services in a given year: that is, what percentage of officers on the rolls were Indians? The other measure is *fresh recruitment*: what percentage of officers recruited each year were Indians?

In the following analysis, I will use two reference points. The first is the government's *recruitment* target of 75 per cent Indians: which, as we saw, was applied to both state and company railways. The second is the 50 per cent *composition* target. Here I make the assumption that, as in the case of the recruitment target, the Lee Commission's original formulation of 'recruited in India' was taken to refer to Indian officers, that is, the long-term goal was to reach a stage when 50 per cent of Superior Service officers would be Indian. This is supported by the fact that the staff and recruitment figures published in official documents under the head of 'Indianization' did not show place of recruitment. For the same reason it is also a viable benchmark in this analysis. Finally, following Railway Board reports, I take Indianization to mean the increasing presence of all statutory Indians: that is, I include Anglo-Indians and Domiciled Europeans in calculations of the Indian share in the Superior Services.

The composition around the time the Lee Commission's recommendations were made is illustrated by Table 4.3, showing the percentage share of Indian Superior officers by department in the state railways and the company-run railways in 1925. In these figures state railways include the newly nationalized (in January 1925) East India Railway, but not the Great Indian Peninsula Railway, which was nationalized later that year.

The overall proportion of Indians in Superior positions was low: 21.49 per cent for all railways taken together. A 50 per cent Indian cadre was clearly a long way away in 1925. Of individual departments, the Agency, which constituted the managerial jobs, was less than one-fifth Indian, while the mechanical engineering jobs (Locomotive, Carriage, and Wagon) were almost entirely European (only 5.09 per cent of its officers were Indian). The Indian proportions in Engineering and Traffic were comparable, around one-quarter in each case. Finally, it is evident that on the whole company-run railways had a lower proportion of Indians than the state railways.

With these as the base figures, the pattern of recruitment clearly had to be heavily in favour of Indian officers for the next several years

Table 4.3   Composition: Extent of Indianization, Railway Superior Service officers, as of 1 April 1925

	Department	Column 1 State Railways, numbers (Indians/Total)	Column 2 State Railways, Indians as %	Column 3 Company-run Railways, numbers (Indians/Total)	Column 4 Company-run Railways, Indians as %	Column 5 Overall, numbers (Indians/Total)	Column 6 Overall, Indians as %
1	Agency	4/23	17.39	7/40	17.50	11/63	17.46
2	Engineering	100/303	33.00	72/435	16.55	172/738	23.31
3	Traffic	65/225	28.89	56/274	20.44	121/499	24.25
4	Locomotive and Carriage and Wagon (LCW)	7/109	6.42	7/166	4.22	14/275	5.09
5	Stores	8/32	25.00	5/43	11.63	13/75	17.33
6	Other Departments	45/110	40.91	39/171	22.81	84/281	29.89
7	Total	229/802	28.55	186/1129	16.47	415/1931	21.49

Source: Based on Appendix G, *Report by the Railway Board on Indian Railways 1924–25*, vol. I (Calcutta: Government of India Central Publication Branch, 1925), pp. 106–7.

Note 1: Of the 1,931 officers (Column 5), 87 were Anglo-Indians (4.51 per cent). These 87 are included in the total of 415 Indian officers.

Note 2: Figures in this table do not include the Nizam's Guaranteed State Railway (Hyderabad) and the Jodhpur Railway, both of them based in princely states.

in order to approach a 50–50 composition of Indians and Europeans. Table 4.4 shows the pattern adopted in fresh recruitment over the 1920s and 1930s.

The figures in Table 4.4 show that on the state railways until the mid-1930s, the percentage of Indians recruited as Superior officers was often in the range of 60 to 70 per cent (Column 6). Although this was a considerable proportion, it fell consistently short of the 75 per cent target. Although the data is not exhaustive, it appears that the highest percentage of Indians were recruited to the Traffic Department (Column 3), which fits with the government's focus on providing training in this area during these years (these measures are discussed later). The company-run railways, which had accepted the Lee Commission's recommendations, fell well below the 75 per cent mark, missing it by a larger margin than the state railways did (Column 8). Sometimes, as in 1933–4 during the Depression, no officers were recruited to the state railways. The last row, for the year 1941–2, shows a sudden jump to almost 100 per cent recruitment of Indians: this is because recruitment in Europe was suspended at the start of World War II.

To put these numbers in perspective, it must be emphasized that recruitment in terms of absolute numbers was very modest throughout the 1920s and 1930s (see Columns 7 and 9). Clearly, then, fresh recruitment could only have a limited impact on changing the racial composition of the officer ranks.

Table 4.5 shows the composition of the Superior establishments of the railways as of 31 March 1934. Ten years after the Lee Commission's recommendations, the overall share of Indians in the Superior Services had risen from 21.49 per cent to 37.85 per cent. This suggests a significant advance in Indianization, although the result was still considerably lower than the 50–50 composition that had been aimed at in the early 1920s: that target would only be met some years later (discussed later). The Indian share of particular departments (Engineering on the company railways, Mechanical Engineering on all railways) was still notably low. In Engineering, too, the increase in the Indian share was at least partly due to the fall in the number of European officers by 202.[103]

[103] Their number was 738 − 172 = 566 in 1925 (Table 4.3); and 600 − 236 = 364 in 1934 (Table 4.5). Thus, it fell by 566 − 364 = 202. See 'Overall numbers' column in each table.

Table 4.4 Recruitment: Indianization—Fresh recruitment to the Railway Superior Service, 1921–41

Column 1	Column 2	Column 3	Column 4	Column 5	Column 6	Column 7	Column 8	Column 9
YEAR	STATE RAILWAYS						COMPANY RAILWAYS	
	Engineering*	Traffic/Transp. (Commercial)*	Mech./Transp. (Power)*	Other*	Overall Superior Posts*	Nos. recruited (Indians/Total)	Overall Superior Posts*	Nos. recruited (Indians/Total)
1921–2	66.7	100.0	–	11.1	65.6	21/32	–	–
1922–3	60.0	100.0	–	23.1	47.8	11/23	–	–
1923–4	71.4	No recruitment	–	14.3	52.4	11/21	–	–
1925–6	–	–	–	–	32.0	–	–	–
1926–7	–	–	–	–	62.0	–	–	–
1927–8	78.0	36.0 (combined)		57.0	68.0	–	49.0	–
1928–9	–	–	–	–	70.0	–	–	–
1929–30	71.0	80.0	73.0	67.0	71.0	–	49.2	–
1932–3	–	–	–	–	63.6	7/11	42.3	11/26
1933–4	No recruitment						59.3	16/27
1934–5	–	–	–	–	61.9	13/21	66.7	26/39
1936–7	66.7	75.0	100.0	100.0	85.7	–	69.6	–
1941–2	100.0	87.5	100.0	100.0	95.5	21/22	97.3	36/37

Figures in columns marked (*) show the percentage share of Indians among fresh recruits in a given year.

Source: Compiled from Railway Board's Administration Reports and *Statement of Moral and Material Progress*, various years.

Note: Empty cells indicate data not available/not found.

Table 4.5 Composition: Extent of Indianization, Railway Superior Service officers, as of 31 March 1934

Department	Column 1 State Railways, numbers (Indians/ Total)	Column 2 State Railways, Indians as %	Column 3 Company-run Railways, numbers (Indians/ Total)	Column 4 Company-run Railways, Indians as %	Column 5 Railway Board & misc. officers, numbers (Indians/Total)	Column 6 Railway Board & misc. officers, Indians as %	Column 7 Overall, numbers (Indians/ Total)	Column 8 Overall, Indians as %
1 Agency	11/37	29.73	9/35	25.71	0/1	0.00	20/73	27.40
2 Accounts	37/51	72.55	28/59	47.46	4/4	100.00	69/114	60.53
3 Engineering	151/335	45.07	82/254	32.28	3/11	27.27	236/600	39.33
4 Transportation	101/247	40.89	81/183	44.26	5/11	45.45	187/441	42.40
5 Commercial*	19/37	51.35	6/13	46.15	0/0	0.00	25/50	50.00
4+5 Transp + Comm (=Traffic)	120/284	42.25	87/196	44.39	5/11	45.45	212/491	43.18
6 Mech. Engg (=LCW)	45/184	24.46	21/133	15.79	0/4	0.00	66/321	20.56
7 Stores	17/43	39.53	11/29	37.93	0/0	0.00	28/72	38.89
8 Other	38/91	41.76	23/67	34.33	12/31	38.71	73/189	38.62
9 Total	419/1025	40.88	261/773	33.76	24/62	38.71	704/1860	37.85

Source: Based on Appendix F, Report by the Railway Board on Indian Railways for 1933–4, Vol. I (Delhi: Manager of Publications, 1934), pp. 103–6.

Notes: 1. Of the total of 1,860 (Column 7), Anglo-Indians and Domiciled Europeans accounted for 125 officers (6.72 per cent). These 125 are included in the total of 704 Indian officers.

2. Figures in this table do not include the Nizam's Guaranteed State Railway (Hyderabad) and the Jodhpur Railway, both of them based in princely states.

* Figures for Commercial are included under Transportation for some Railways. Figures in this row refer to railways where separate figures are given for the Commercial Department.

In the Depression-hit early 1930s, the Railway Board began a policy of retrenchment. In 1932, for instance, it asked the state railways to reduce its officer cadre substantially, dissolved a number of posts on the company lines, and ended the services of several temporary engineers.[104] In these years the strength of the European officers in particular fell. Daniel Headrick has suggested that Depression-era conditions made the Indian railways an unattractive destination for potential British officers.[105] In addition, some Europeans may have taken voluntary retirement, though it has not been possible to ascertain this. At any rate Tables 4.3 and 4.5 show that between 1925 and 1934, when the overall number of Superior officers dropped by seventy-one (from 1,931 to 1,860), the drop in the number of Europeans works out to the much larger figure of 360 (whereas the net increase in Indian officers was 289). Thus, although the *recruitment* percentages followed (Table 4.4) contributed to the relative Indianization of the railways, the departure of European officers evidently contributed to the rise in Indians' percentage share overall.[106]

Finally, it should be noted that the Anglo-Indians and Domiciled Europeans formed only a minor component (well below 10 per cent) of the Superior Services throughout the 1920s and 1930s, although they did form a substantial proportion of (statutory) Indian officers. In addition, they were predominant among Upper Subordinates, that is, subordinate employees with a monthly salary of Rs 250 and above, or pay scales rising to that level.[107]

---

[104] Natesan, *State Management & Control of Railways*, pp. 369–70. Kerr, in *Engines of Change* (p. 130), also refers to retrenchment in the 1930s.

[105] Headrick, *Tentacles of Progress*, p. 343.

[106] Headrick (*Tentacles of Progress*, p. 343) suggests that Depression conditions and retrenchment boosted Indianization in these years. However, this is a general observation and is not based on Superior Service employment statistics.

[107] The percentage shares among Upper Subordinates in 1925 were: Europeans 35.84 per cent; Anglo-Indians 37.61 per cent; other Indians 26.55 per cent. In 1934 they were: Europeans 19.59 per cent; Anglo-Indians and Domiciled Europeans 42 per cent; other Indians 38.41 per cent. Appendix G in *Report by the Railway Board on Indian Railways for 1924–25*, vol. I and Appendix F in *Report by the Railway Board on Indian Railways for 1933–34*, vol. I (Delhi: Manager of Publications, 1934). These figures are in general agreement with Daniel Headrick's observation that '[d]uring and after World War I, the railways replaced many of their midlevel imported European personnel with resident Europeans and Eurasians' (Headrick, *Tentacles of Progress*, p. 342).

Figures 4.1 and 4.2, based on Tables 4.3 and 4.5, summarize the statistics presented on Superior Officers.

The figures show that the Indian presence grew in all departments across all railways; that their overall share on the state lines was always higher than on the company lines; that the most dramatic increase came in the company-run railways' Traffic Departments; and that while the Indian share in Mechanical Engineering increased, it was still the least Indianized of all departments in 1934. The charts also show that despite these developments, no department on either type of railway had attained a 50–50 composition of Indian and European Superior officers as of 1934.

In fact, it was only towards the end of the 1930s that Indians exceeded the 50 per cent mark in the Superior Services. The decisive

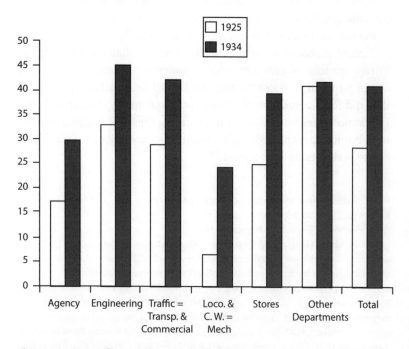

Figure 4.1   State railways: Indians by percentage in each department of the Superior Services, 1925 and 1934

(Y axis shows Indians' percentage share of appointments in each department)

Source: Based on Tables 4.3 and 4.5.

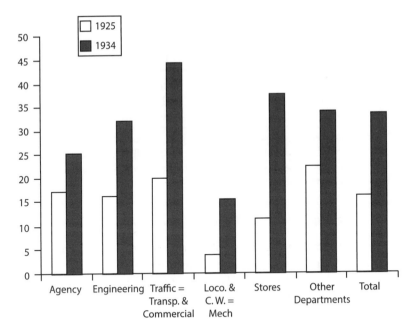

Figure 4.2 Company-run railways: Indians by percentage in each department of the Superior Services, 1925 and 1934
(Y axis shows Indians' percentage share of appointments in each department)
Source: Based on Tables 4.3 and 4.5.

increase in the Indian share came after the outbreak of World War II, when recruitment in Europe ceased, and many European officers left India to serve in the war.[108] Thus, the departure of Europeans was a critical factor once again. The numbers under each department for these years have not been found, but the overall figures are shown in Table 4.6.

Thus, the achievement of a 50–50 composition a full decade and a half after the Lee Commission came about not only with the recruitment of more Indians, but also with the intervention of external factors (the Depression and World War II) that caused the exit of European officers from railway service in India.

[108] See Kerr, *Engines of Change*, pp. 119, 130–1.

Table 4.6 Composition: Percentage share of Indians among gazetted railway officers in the early years of World War II

	1941	1942
State railways	60.82	64.14
Company-run railways	51.78	57.26
Overall	56.69	60.95

Source: *Report by the Railway Board on Indian Railways for 1941–2*, vol. I, p. 30.

Note: These numbers include the Superior Service and a small 'Lower Gazetted Service', an intermediate officer level that was small enough that the Superior Service taken in isolation would have almost the same percentage figures.

Training Indians: Non-Uniform Measures across Departments

The high rate of recruitment of Indians to the Traffic Department (Table 4.4) and the relatively high Indian presence in that department in 1934 (Figures 4.1 and 4.2) point to a conscious effort to seek and train potential Traffic officers in India. This is corroborated by the qualitative sections of the Railway Board's annual reports in this period, which almost invariably mentioned facilities and schemes instituted for training current or potential Traffic officers. These sources also mention training schemes for Locomotive and Carriage officers, though to a lesser extent.[109]

From the early twentieth century, individual railways had various types of apprentice schemes to train employees for work. The various lines had permanent or temporary railway schools in different parts of the country, as in Asansol and Jamalpur (on the East Indian Railway), Lyallpur and Walton, Lahore Cantonment (on the North-Western Railway) and Bina (on the Great Indian Peninsula Railway). These 'area schools' catered mainly to subordinates, with courses for Switchmen, Assistant Station Masters, Traffic Inspectors, and Rolling Stock operators. It appears that they obtained their students mainly from among the sons of railway employees with a basic school education.[110]

---

[109] The following account of training facilities is based on issues of the *Statement Exhibiting Moral and Material Progress* and on the annual reports of the Railway Board, for various years between 1920 and 1940.

[110] For instance, G. L. Colvin, Agent of the East India Railway (EIR), placed an advertisement in the *Indian Railway Gazette*, January 1923, p. 5, calling for European, Anglo-Indian, and Indian candidates for subordinate posts in the Engineering

In 1925 a school was set up in the northern town of Chandausi to coordinate the activities of the other schools, and to train probationary Traffic officers. After a few years, though, it confined itself to training subordinates, prompting the inauguration in 1930 of a Railway Staff College at Dehra Dun in the Himalayan foothills. The Railway Staff College was to cater primarily to Superior officers in the Transportation (Traffic and Mechanical) Department. Officers selected in India would spend a total of five months at the Railway Staff College as part of their training. In a profile of the new Staff College, the *Indian Railway Gazette* wrote:

> The training is ... practical and theoretical, and is in pursuance of the accepted policy of the Government of India to increase the number of Indians employed in the higher grades of railway work. Young Indian gentlemen [this probably refers to university graduates] who wish to adopt railway Transportation as their profession, have the satisfaction of knowing that, if they succeed in passing the initial examination, they can look forward to training in their profession probably unequalled anywhere else in the world.[111]

As it happened, the need to cut costs in the Depression years caused the Staff College to be shut in 1932. The Railway Board asked individual railways to prepare programmes of lectures for their officers, and went back to relying on the area schools on the various lines.

We have seen that Superior officers in the Locomotive and Carriage (Mechanical Engineering) departments were overwhelmingly British: recruitment depended not on college training but on practical skills gained in railway workshops, and the Indian railway workshops did not have systematic training facilities. In the 1920s attempts were made to address this deficiency. Among the more concrete measures taken was the East India Railway's (EIR) Special Class Apprentice

---

Department of the EIR. Candidates had to be aged between sixteen to twenty and have passed the Junior Cambridge or equivalent exam (if they were European or Anglo-Indian) or the Matriculation Exam in the First Division (or equivalent) if they were Indian. Sons of EIR employees and graduates of the school in the railway town of Jamalpur would be given preference.

[111] 'Railway Staff College, Dehra Dun: A Further Step in the Progress of India's Railways', *Indian Railway Gazette*, January 1930, pp. 13–15.

(SCA) programme for candidates from across the country, in operation from the mid-1920s (around the time the EIR was nationalized).
It was based in the Jamalpur workshop and the adjoining technical
school. A description of this programme is found in the memoirs of
Duvur Venkatrama Reddy, a Mechanical Engineering officer in the
EIR (and, after Independence, Indian Railways). Reddy, a student
at Presidency College in Madras, was among twelve matriculates
selected from all over India for the programme in 1930. He underwent four years' training at Jamalpur, followed by two years in England in railway workshops at Ashford and Eastleigh, before returning
to join the EIR's Superior Service.[112]

On the basis of government reports, then, we can say that most
schemes put in place to train Indians focused on instruction in (a)
Superior functions in Traffic, (b) Superior functions in Mechanical
Engineering, and (c) Subordinate functions. In contrast, there are
virtually no references in this official literature to any schemes to
increase opportunities for Indian officers in the Civil Engineering
Department, or for the specialist posts of bridge, signal, and electrical engineer. Only one exception was found, a scheme proposed
in 1926–7, under which the Railway Board was to select seventeen
Indian graduate engineers (nine from British colleges and eight from
Indian colleges) every year for a year's 'practical training on railway
construction', but with no guarantee of permanent employment at
the end of it. This scheme, proposed during a particular spell of fresh
construction, was a one-off development; indeed it is not clear if it
materialized or how long it lasted.[113]

The absence of greater discussion on providing practical training
for civil engineers is remarkable, given that such training was considered an essential requirement before a graduate engineer could
be employed in the railways. In the face of official silence, it is difficult to establish with certainty the reason for this. However, a clue
is provided by some of the questions put to the Railway Board in the
Legislative Assembly, which were concerned specifically with Indianization in engineering posts, as opposed to the Superior Services as
a whole. A telling example is a question put by Lala Duni Chand in

[112] Reddy, *Inside Story of the Indian Railways*.
[113] *Statement Exhibiting Moral and Material Progress*, 1926–7, pp. 173–4.

1926. He asked if it was true that the Railway Board had 'virtually refused to carry out' the Indianization of higher posts; that Indian graduates, even after training, were deemed unfit for higher appointments; that Indians with qualifications from the Indian engineering colleges were 'invariably refused appointments in the Railway Engineering Service'; and that 'engineers turned out by the Engineering Colleges of the United Kingdom are appointed railway engineers on high salaries while equally qualified and far less expensive Indians are not appointed'. The charges were denied brusquely, but the question served to demonstrate politicians' scepticism about the authorities' commitment to Indianization—especially with regard to the Civil Engineering Department, which, as we saw earlier, was the most prestigious in the railway hierarchy.[114] This fact, combined with the attitudes discussed earlier, which viewed Indians as inefficient and their loyalty to the state as questionable, probably made it important, in the eyes of the Railway Board, to limit the pace of Indianization in the engineering establishment. Put another way, it is likely that greater measures were taken to Indianize the middle and lower levels of the railway bureaucracy than the highest.

Indeed, despite the various steps towards Indianization in some areas, the highest echelons of the railways' Superior Services were marked by continuity over our period: European dominance was still their most visible feature. There were few Indians in the higher ranks of the railway Engineering Departments and their upper management. As late as 1940, there was only one Indian among the Chief Engineers and General Managers on the state railways (L. P. Misra, a Roorkee engineer, General Manager of the Eastern Bengal Railway).[115] By contrast, Indian Chief Engineers appeared earlier in the princely states, whose internal policies were largely independent of the colonial government. For instance, in 1930 the Cutch State Railway appointed as Manager and Engineer-in-Chief S. K. Kothari, an Indian engineer.[116]

[114] *Extracts from the Debates in the Indian Legislature on Railway Matters: Delhi Session—January to March 1926*, p. 79.

[115] *India Office List* for 1940; 'Sir Lakshmi Pati Misra' in 'Luminaries', available at http://www.iitr.ac.in/institute/pages/Heritage+Luminaries.html (accessed 22 July 2011).

[116] See portrait of Kothari and caption in *Indian Railway Gazette*, January 1930, p. 3.

The Gaekwar's Baroda State Railway had only a handful of engineers listed (all of them Indian) in 1931 and 1932, many of them affiliated to the Institution of Engineers (India).[117] Other features of the railway bureaucracy remained unchanged through the 1920s and 1930s. A perusal of the *India Office Lists* for various years shows, for the state railways, the continued presence of Royal Engineers throughout this period. It also shows that Indian technical specialists, that is, bridge, signal, and electrical engineers, were few and far between.

Committees formed to discuss points of cooperation between the various railway systems (whose lines were interlinked) were also staffed primarily by British officers. The Indian Railways Conference Association (IRCA) was formed in 1902 'to frame rules and regulations for booking of traffic and interchange of trains between railways'.[118] This body, made up of the top management (most of them engineers) of the various lines, was predominantly European in our period, reflecting the fact that Indians had not yet climbed to the top rungs of the railway hierarchies.[119] There were no Indians among the delegates of the IRCA's 1920 meeting, who included Major-General Sir H. F. E. Freeland, Agent, Bombay Baroda and Central India Railway (presiding); C. D. M. Hindley, Agent of the East Indian Railway; and F. A. Hadow, Agent of the North-Western Railway.[120] Sixteen years later, the IRCA was still a British-dominated body, as indicated by its office-bearers. The President was J. C. Highet, AMICE, Agent of the North-Western Railway, Lahore; the General Secretary

[117] *Thacker's Indian Directory* for 1931 and 1932. On the Institution of Engineers (India), a professional society, see Chapter 2 of this book.

[118] 'Indian Railways: Indian Railway Conference Association: History', available at http://www.indianrailways.gov.in/railwayboard/uploads/directorate/IRCA/index.jsp (accessed 17 July 2012).

[119] Examples of topics discussed include: 'Preparation of a "Code" or standard list of electrical apparatus, fittings and material in regular demand on railways in India' and 'Standardisation of voltages (pressures) A.C. and D.C. to be adopted in railway workshops for motors, lighting and portable tools'. Indian Railways Conference Association 1928, *Agenda and Proceedings of the Electrical Section: Meeting No. 1* (place and date of publication not given), p. i. British Library Shelfmark: W 4026.

[120] Indian Railway Conference Association, *Proceedings of the Conference of Railway Delegates Assembled at Simla, October 1920* (Simla: Government Central Press, 1920) [APAC: IOR/V/25/720/42].

B. Lawrence, MICE, Delhi; and the Deputy General Secretary S. S. Stubs, BA Mechanical Science (Cambridge), of Delhi.[121]

Finally, the Railway Board itself exhibited a similar trend. Indians rarely rose to the top positions. The lists of its office-bearers in three sample years (1934, 1937, and 1942) reveal that the offices of Chief Commissioner, Railways, and Director of technical departments like Civil Engineering and Deputy Directors of Mechanical Engineering were held almost without exception by Europeans. The highest post attained by an Indian was Financial Commissioner (Sir Raghavendra Rau and T. S. Sankara Aiyar, the latter an engineer by training).[122]

***

The senior staff of the railways in interwar India formed a technical bureaucracy in which engineers of various types played an important role. Of these 'Superior Service' officers (on state and company railways taken together) nearly 80 per cent were Europeans at the start of the 1920s. In response to demands by increasingly empowered nationalist politicians, the government agreed on a policy of Indianizing this cadre, setting itself a dual target: achieving a 50 per cent Indian share in the cadre (within an unspecified time frame), and recruiting fresh officers in the ratio of three Indians to every European.

However, the actual ratio in which recruitment was carried out in the following decade consistently fell short of this formula of 75 per

---

[121] 'Directory of Railway Officials', *Indian Railway Gazette*, January 1936, p. iii. APAC: SM 150 (Microfilm), Box: 1936–1937.

[122] Appendix H: 'Officers of the Railway Department (Railway Board) and Attached Offices, on 31st March 1934', *Report by the Railway Board on Indian Railways for 1933–4*, vol. I (Delhi: Manager of Publications, 1934), p. 114; Appendix H: 'Officers of the Railway Department (Railway Board) and Attached Offices, on 31st March 1937', *Report by the Railway Board on Indian Railways for 1936–7*, vol. I (Delhi: Manager of Publications, 1937), pp. 152–3; Appendix A: 'Officers of the Railway Department (Railway Board) and Attached Offices, on 31st March 1942', *Report by the Railway Board on Indian Railways for 1941–2*, vol. I (Delhi: Manager of Publications, 1943); p. 36. One P. R. Rau is listed as Financial Commissioner for 1933–4, and is most probably the same person as Sir Raghavendra Rau (Financial Commissioner 1936–7).

cent Indians (it was more in the range of 60 to 70 per cent). It was not until the late 1930s that the targeted 50–50 composition of Indians and Europeans was reached. The retrenchment or departure of European officers during the Depression and after the outbreak of World War II was a significant factor in this change in composition, the contribution of which was as important as fresh recruitment of Indian officers. Further, I have demonstrated that the government's Indianization strategy varied across railway departments. While new facilities were put in place for the training and recruitment of Indians in the Traffic and Mechanical Engineering Departments, and for subordinate posts, the traditionally pre-eminent Civil Engineering Department (and specialist positions such as bridge and signal engineering) saw no fresh measures, and a lower Indian share among those recruited. The most likely explanation for this is that Civil Engineering was considered the department with the greatest responsibility (most of the top management were drawn from this department), and hence the one for which Indians were least fitted, given the prevailing conceptions of their character and abilities.

In general the government, while making several concessions to the political currents of the time and acceding to Indianization in principle, considered the railways a vital part of the colonial state, and was careful to avoid changing the character of their senior staff drastically. The railways' important role in internal and external security and their susceptibility to sabotage in case of mass unrest made the presence of 'loyal' (European) engineers in them crucial. There was also a pervasive feeling among influential members of the railway community that the railways must be run like profit-making businesses, and that there was little room for recruiting practices that might lead to a lowering of 'efficiency'—that is, Indianization. Thus, security concerns, coupled with fears of inefficiency in the running of the service and a negative view of Indians' aptitude and capabilities, made the authorities cautious in undertaking the Indianization of engineering and other technical jobs in the railways.

The cases of the PWD and the railways show that the two largest traditional employers of engineers in colonial India were gradually, if reluctantly, bending to the winds of change in allowing Indian engineers a more prominent role in the interwar years. In both cases colonial officialdom had to contend with nationalist demands and the changing structure of the Indian polity. The dynamics of Indianization, however, were somewhat different in private industry, which was fast emerging as an important employer of engineers in the subcontinent. The next chapter explores the experience of industrial engineers.

CHAPTER FIVE

Beyond Empire and Nation

The Technical Experts of the Tata Steel Works

In the interwar period Indian business families (such as the Tatas, Birlas, Walchands, and Kirloskars) began to invest heavily in large-scale industry, challenging European-owned firms for the first time.[1] Of these Indian-owned enterprises, the Tata Iron and Steel Company (TISCO) was exceptionally large and successful. It was the first major producer of steel in India, and achieved a near-monopoly in the domestic market. Its success was instrumental in the creation and operation of several ancillary industries that consumed steel.

Economic historians have argued that TISCO was an exceptionally successful case among large-scale industrial enterprises in interwar India, in that it overcame many of the constraints on industrialization under the colonial government. The growth of most industries, they argue, was limited by factors such as the government's fiscal policy, inadequate protection of Indian industry, lack of domestic demand, scarcity of capital, lack of technological know-how, and the risk-averse behaviour of India-based industrialists.[2] The success of

[1] Dietmar Rothermund, *An Economic History of India: From Pre-Colonial Times to 1991*, 2nd edn (London: Routledge, 1993), p. 92; Rajat K. Ray, *Industrialization in India: Growth and Conflict in the Private Corporate Sector 1914–47* (New Delhi: Oxford University Press, 1982 [1979]), p. 5; B. R. Tomlinson, *The Economy of Modern India, 1860–1970*, The New Cambridge History of India, III.3 (Cambridge: Cambridge University Press, 1996 [1993]), p. 143.

[2] Rothermund, *An Economic History of India*, pp. 61–5; Ray, *Industrialization in India*, pp. 3, 81, and ch. 4; A. K. Bagchi's argument as described in Tirthankar Roy, *The*

TISCO is attributed variously to the vision and foresight of the company's founders, the management's good relations with the colonial government and its resultant obtaining of economic protection, capital investments made after World War I, and effective sales strategies.[3] Yet the literature does not pay sufficient attention to another very important factor: TISCO's strategies in recruiting and training its engineering personnel. Similarly, the historiography of science and technology tells us little about the experience of scientists and engineers working in large-scale industrial firms, focusing as it does on intellectuals' debates on the role of science-based industry in the future nation.[4] An exception is Daniel Headrick's brief account of the creation of TISCO and the Indianization of its technical personnel.[5] Headrick uses his account to illustrate his larger argument that 'technology transfer' in colonial India (the introduction of particular technologies, the required machinery, and the associated expertise and technical culture from the metropolis into the colony) was of a limited nature.[6] This chapter, by contrast, takes the industrial experts themselves—their backgrounds, skills, and characteristics—as its primary subject.

In addition to published memoirs and company newsletters, this chapter makes extensive use of official correspondence, memoranda, and records in the Tata Steel Archives to identify several technical experts of TISCO in the interwar period and study them in detail.

Economic History of India 1857–1947, 2nd edn (New Delhi: Oxford University Press, 2006), p. 263. Although the literature emphasizes the limits of industrial growth, it does show that there was an appreciable growth in large-scale industry over the interwar period (see Chapter 1).

[3] See Ray, *Industrialization in India*, pp. 74–93; Chikayoshi Nomura, 'Selling Steel in the 1920s: TISCO in a Period of Transition', *Indian Economic and Social History Review*, vol. 48, no. 1 (2011): 83–116.

[4] For example, Shiv Visvanathan, *Organizing for Science: The Making of an Industrial Research Laboratory* (Delhi: Oxford University Press, 1985); Gyan Prakash, *Another Reason: Science and the Imagination of Modern India* (Princeton: Princeton University Press, 1999), chapter 6; Pratik Chakrabarti, *Western Science in Modern India: Metropolitan Methods, Colonial Practices* (Delhi: Permanent Black, 2004).

[5] Daniel Headrick, *The Tentacles of Progress: Technology Transfer in the Age of Imperialism, 1850–1940* (New York and Oxford: Oxford University Press, 1988), pp. 285–94, 371–4.

[6] See Headrick, *The Tentacles of Progress*, p. 9 for his definition of technology transfer.

It shows that while the role of British engineers—and of technical instructors schooled in the university and industries of Sheffield—was significant, the technical work of TISCO was not led or directed by British expertise (unlike in the case of the public works and railways). Rather, Americans and American-trained Indians were the most important of the multinational group of experts that ran the technical operations of this pre-eminent industrial enterprise. The existence of foreign experts is recognized in the existing literature. However, their training, the nature of their expertise, and the culture of engineering they brought to TISCO have not been studied before. Here, I will show that these superintendents and managers, many of whom were schooled in the steel works of their home countries, shaped a working culture that laid great store by practical experience, physical fitness, and presence of mind on the shop floor.

These findings will be linked to the ongoing analysis of Indianization. Within TISCO, this development was systematized in the inter-war period through the Tatas' Jamshedpur Technical Institute (JTI, est. 1921), a pioneering step in industrial education in the country. Several accounts of TISCO mention the Institute and its role in Indianization,[7] but none explores at any length the Institute's functioning (such as the staff, curriculum, funding, selection of students, and the educational context in which it was established and operated). Focusing on these aspects, this chapter argues, as Chapters 3 and 4 did for public works and the railways, that a level of continuity in the work culture was maintained while pursuing Indianization. Indians trained in the Institute were expected to combine a mastery of the theory of steel manufacture with intensive practical learning in the TISCO works itself. Physical fitness and industriousness remained important, being key parameters in the selection of students for the Institute, and thereafter for jobs in the works. This remained true even when the Institute's main programme evolved into a more specialized one designed for graduates in metallurgy and engineering.

---

[7] For example, Ray, *Industrialization in India*, p. 91; Headrick, *Tentacles of Progress*, p. 373; and Hiruyoki Oba and Hrushikesh Panda (eds), *Industrial Development and Technology Absorption in the Indian Steel Industry: Study of TISCO with Reference to Yawata—A Steel Plant of Nippon Steel Corporation in Japan* (Mumbai: Allied Publishers, 2005), chapter 6.

Despite these similarities, the pattern of Indianization within TISCO was markedly different from that in government technical services. In this private company owned entirely by Indian capital, Indianization was mainly an internally driven process. Political demands for Indianization played a minor and indirect role here, whereas in the case of public works and railways, public opinion and the continuous pressure applied by Indian legislators were crucial drivers of Indianization. Further, whereas most Indian technical experts who entered government services were educated at the state's engineering colleges (which had the patronage of PWD appointments) or trained at government-run railway workshops and technical schools, the role of the colonial education system in the Indianization of TISCO was small—for this the company relied on its own institute, the JTI, and on Indians trained abroad.

MULTINATIONAL EXPERTS AND THE CULTURE OF STEEL-MAKING

Although the Tata Iron and Steel Company was formally registered in 1907, the first steps towards its establishment had already been taken at the turn of the twentieth century by Jamsetji Nusserwanji Tata, a successful owner of textile mills in Bombay and Nagpur. According to his biographer, J. N. Tata had long nurtured the ambition of setting up an iron and steel works in India, where no producer of steel existed. By the end of the nineteenth century, preliminary reports had been published indicating iron and coal deposits in India, and in 1899 the government relaxed its restrictions on mining by private agencies. Tata then secured prospecting licences for some districts in central India and had his representatives explore the area for ore.[8]

From this stage through to the early years of the plant's functioning, the technical work of the company was carried out by an international group of experts under the direction of Americans. This expertise was embodied in several types of personnel: those who carried out the initial prospecting; the construction engineers who built the

[8] R. M. Lala, *For the Love of India: The Life and Times of Jamsetji Tata* (New Delhi: Penguin/Portfolio, 2006 [2004]), chapter 18. Although a company called the Bengal Iron Works was already functioning, it had so far manufactured only iron, not steel (Headrick, *Tentacles of Progress*, pp. 282–4).

plant itself; the technical superintendents and managers who ran the various production departments; and the experts in charge of the rest of the plant and the township in which it was located.

The role of American construction engineers began soon after Tata had acquired his prospecting licenses. In the early 1900s, he travelled to the USA in search of expert assistance. In doing this he not only acknowledged the contemporary success of America's steel industry, but also took the then unusual step of venturing beyond the British Empire. While in America, Jamsetji 'studied coking processes at Birmingham, Alabama, visited the world's largest ore market at Cleveland, and in Pittsburgh met the foremost metallurgical consultant, Julian Kennedy'. Kennedy agreed to build a works in India, provided a detailed survey of materials was conducted. He recommended for the survey the New York consulting engineer C. P. Perin.[9]

The grandson of an engineer, Charles Page Perin (b. 1861) had an A.B. from Harvard College, and had also studied at the Écoles des Mines in Paris. Working his way up as an engineer in the Carnegie Steel Company and in steel plants in Alabama, Kentucky, and Tennessee, Perin set up as a consulting engineer in 1900. He was credited with the building of coke plants, whole industrial towns in Virginia, and an electrolytic iron works at the Niagara Falls. In the course of his career he worked in various countries including China, Spain, South Africa, and Russia.[10] It was in 1902 that Jamsetji met him upon Kennedy's suggestion. Perin accepted the role of chief consulting engineer to Tata's putative steel works (his firm continued to be consulting engineers to the Tatas' steel company until 1936).[11]

Over the next few years Perin's partner C. M. Weld successfully led explorations in North-Central India, culminating in the selection of a site for the steel plant: the village of Sakchi, less than 200 miles west

[9] R. M. Lala, *The Creation of Wealth: The Tata Story*, paperback edition (Bombay: IBH, 1981), chapters I and II. The quote is from p. 20.

[10] 'Dr. Charles Perin, Engineer, 75, Dies', *New York Times*, 17 February 1937, p. 21.

[11] Lala, *Creation of Wealth*, p. 20; R. M. Lala, *The Romance of Tata Steel* (New Delhi: Penguin/Viking, 2007), p. 62. For the date of Jamsetji's trip and the duration of Perin's association with the Tatas, see '"C. P."' in *TISCO Review*, April 1937, p. 255. Several issues of *TISCO Review* and *TISCO News* were consulted at the Tata Steel Archives, Jamshedpur (hereinafter referred to as TSA), and a few at the Tata Central Archives, Pune.

of Calcutta.[12] Although Jamsetji died in 1904,[13] the enterprise continued under his son Dorabji. In 1907 the Tata Iron and Steel Company
was registered, a sum of Rs 23.2 million having been raised from
investors in India. Tata and Sons (later Tata Sons) of Bombay held a
share of 11 per cent in the new company and were appointed TISCO's
managing agents.[14]

The American engineers of Julian Kennedy's and Perin's firms
led the building of the new plant at Sakchi (later renamed Jamshedpur after Jamsetji Tata). Axel Sahlin, from Julian Kennedy, Sahlin
and Company, arrived in February 1908. He was accompanied by
W. O. Renkin (c. 1875–1943), who was appointed resident construction engineer. Renkin was a Pittsburgh native who, in his lifetime,
held high positions in various companies such as the Quigley Fuel
Company (New York), the A. M. Byers Company (Pittsburgh), and
the Coke Dry Quenching Equipment Corporation.[15] C. M. Weld, who
had led the prospecting efforts, stayed on to supervise the work until
the arrival of the first in a long line of American General Managers,
R. G. Wells, in January 1909.[16] Wells, an expert in the building of
iron works, had worked previously in Mariopol in imperial Russia and

---

[12] Lala, *Creation of Wealth*, pp. 20–3; C. Minot Weld, 'The Beginnings of the Tata
Iron & Steel Company', *TISCO Review*, November 1933, pp. 2–8. The Tatas had been
led to the Gorumaishini deposits by P. N. Bose, formerly of the Geological Survey
of India and at this time state geologist of Mayurbhanj, the princely state in which
the deposits were located (Jogesh Chandra Bagal, *Pramatha Nath Bose* [New Delhi:
Sushama Sen on behalf of P. N. Bose Centenary Committee, 1955], chapter VIII).

[13] Lala, *Romance of Tata Steel*, p. 13.

[14] Lala, *Creation of Wealth*, pp. 21–4; Weld, 'The Beginnings of the Tata Iron &
Steel Company' ; Verrier Elwin, *The Story of Tata Steel* (Bombay: n.p., 1958) [British
Library Shelfmark W 3092], p. 37; Lala, *Romance of Tata Steel*, p. 17n2; John L. Keenan
[with Lenore Sorsby], *A Steel Man in India* (New York: Duell, Sloan and Pearce, 1943),
p. 34. Rothermund in *Economic History of India* (p. 61) argues that investing in TISCO
afforded wealthy Indians a means of 'showing their patriotism' in a discreet way.
These rich Indians included rulers of the princely states, who were under the ultimate
authority of the colonial government.

[15] Weld, 'The Beginnings of the Tata Iron & Steel Company', p. 5; 'The Late Mr.
W. O. Renkin', obituary appearing in *TISCO Review*, December 1943, p. 242. Sakchi
was renamed at the end of World War I (Lala, *Romance of Tata Steel*, p. 29).

[16] Weld, 'The Beginnings of the Tata Iron & Steel Company', p. 8.

with the Dominion Iron and Steel Company in Sydney, Nova Scotia (Canada).[17]

A number of British and Indian engineers worked with these Americans. In 1908–9 a Scottish mining engineer, McNeil, was drafted in to assist with the mining of ore, while an English engineer, B. B. Willcox, was engaged to assist C. M. Weld until his departure from Sakchi. Srinivasa Rao, an Indian graduate of the Mysore State Geological School, had, along with Weld, Dorabji Tata, and Shapurji Saklatvala (Dorabji's cousin), been a part of the early prospecting expeditions; after his premature death from cholera, another Indian, Vyas Rao, took over the prospecting and geological work. Weld records that the site chosen at Sakchi was surveyed by 'a corps of native engineers' under Willcox.[18] K. R. Godbole, an Indian who had previously worked in the Public Works Department, was made civil engineer responsible for amenities in the township that would be built in Sakchi.[19]

Unsurprisingly, this mixed group of experts had diverse styles of functioning. A contemporary observer, Mrs B. J. M. Cursetjee, referred to '[w]atchful Weld', '[v]igorous, forceful, impressive Perin', and 'tall, broad-shouldered, active' Sahlin, and indicated that there was some friction: 'What a struggle between blustering, bullying Renkin who wanted work pushed at any cost and patient, white-turbaned, methodical, Godbole.'[20] Axel Sahlin found Godbole something of a curiosity, writing: 'the Tata Co have a Civil Engineer, Mr Codbole [sic], who is a Brahmin of high caste. I do not know how many washings it will take him to get clean after he has associated with us an entire day.'[21] These minor clashes notwithstanding, work progressed, and the plant soon became operational, producing its first batches of steel in 1912. The plant was built with a capacity of 72,000 tons of finished steel per year. The machinery installed, which had been purchased in America and Germany, comprised two blast furnaces (capacity 175 tons per day),

[17] F. R. Harris [with Lovat Fraser], *Jamsetji Nusserwanji Tata: A Chronicle of His Life*, 2nd edn (Bombay: Blackie & Son, 1958), p. 203.

[18] Weld, 'The Beginnings of the Tata Iron & Steel Company' (the quoted words are on p. 5); Lala, *Creation of Wealth*, chapter II.

[19] See Lala, *Romance of Tata Steel*, p. 20.

[20] B. J. M. Cursetjee, quoted in R. M. Lala, *Romance of Tata Steel*, p. 20.

[21] Axel Sahlin, 'Personal Impressions of India: Written for His friends by Axel Sahlin: January 15th to April 21st, 1908', p. 24. Booklet consulted at TSA.

four Siemens-Martins open hearth furnaces, a blooming mill, a rail and beam mill, two bar mills, and 180 Coppee coke ovens.[22]

Foreign consultants and technical advisers continued to be important after the inauguration of the plant. When problems were faced with the open hearth furnaces just before World War I, Charles Perin came to Jamshedpur along with Ralph Watson, an open hearth expert whose services were lent to him by the Carnegie Steel Company, and the furnaces were set right.[23] Perin was also a critical figure during the Greater Extensions programme, an expansion of the plant that began in 1916, aiming to raise the works' output by a factor of five.[24] In his New York office, Perin took on two new partners, and employed 300 engineers and draftsmen, sending 700,000 tracings and three million blueprints to India in the years 1917–20.[25] In the interwar years, the company employed as its Technical Advisor a British expert named Richard Mather.[26] Born in 1886, he was educated at Sheffield University, worked at the Ormsby Iron Works in Middlesborough, did metallurgical research for the British War Office in Woolwich, and was Metallurgical Inspector to the Government of India before he joined the Tatas.[27]

Meanwhile, the second major category of technical experts arrived in Sakchi around 1910. These were the superintendents and managers who would run the various production and support departments. We may estimate that they numbered in the range of twenty to thirty (based on Figure 5.1). Initially most of them were foreigners. These experts, in addition to skilled workers from various foreign countries,

[22] 'The Tata Iron and Steel Company Ltd. 1907 to 1934', *TISCO Review*, November 1934, pp. 718ff, here pp. 718, 722; Headrick, *Tentacles of Progress*, p. 291.

[23] Keenan, *Steel Man*, pp. 42–3.

[24] Lala, *Romance of Tata Steel*, pp. 36–7.

[25] Lala, *Romance of Tata Steel*, pp. 37–9; Keenan, *Steel Man*, p. 68.

[26] Mather's designation as given in a letter from TISCO (signature illegible) to S. R. Dongerkeri, Registrar of Bombay University, 30 November 1937. Folder: D-51, Box 51: Private Papers, R. Mather, TSA.

[27] Cutting of entry on Mather in *Illustrated All-India Trade Directory and Who's Who*, 1942. Enclosed with letter from Barque and Company, Lahore (publishers of the *Who's Who*), 20 January 1943, inviting him to make changes if required for the 1943 edition. Folder: D-51, Box 51: Private Papers, R. Mather, TSA.

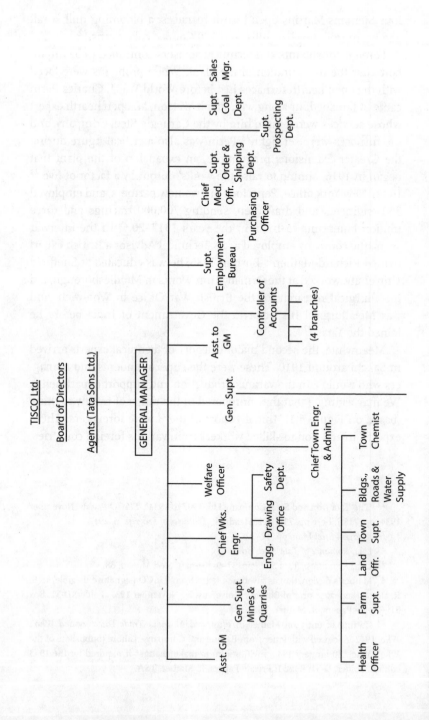

TISCO Ltd.

Board of Directors

Agents (Tata Sons Ltd.)

GENERAL MANAGER

- Asst. GM
  - Gen Supt. Mines & Quarries
  - Chief Wks. Engr.
    - Engg. Drawing Office
    - Safety Dept.
  - Welfare Officer
- Gen. Supt.
  - Chief Town Engr. & Admin.
    - Health Officer
    - Farm Supt.
    - Land Offr.
    - Town Supt.
    - Bldgs., Roads & Water Supply
    - Town Chemist
  - Asst. to GM
    - Controller of Accounts
      - (4 branches)
  - Supt. Employment Bureau
  - Purchasing Officer
  - Chief Med. Offr.
  - Supt. Order & Shipping Dept.
  - Supt. Prospecting Dept.
  - Supt. Coal Dept.
  - Sales Mgr.

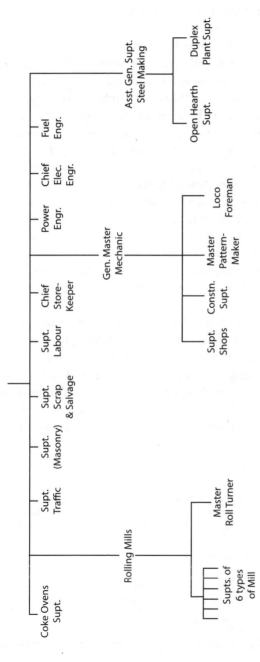

Figure 5.1 Organization structure of Tata Iron and Steel Company (TISCO), interwar period

Source: Abridged from undated diagram titled: 'A Chart showing the Organisation of the Staff, both administrative and departmental with the monthly Expenditure of each Dept.', courtesy Tata Steel Archives, Jamshedpur.

Note: That the chart refers to the interwar period is inferred based on the expenditure figures marked on the original chart. For example, expenditure under General Manager (all or most of which would have been his salary) is marked Rs. 7,500—which is broadly commensurate with the average salary figures in Table 5.1 below (those figures include bonuses).

made up a foreign contingent of about 175 in an overall workforce of around 2,000.[28]

All the foreigners, irrespective of designation, were employed on renewable contracts and referred to as 'covenanted' staff. The usage of the term was similar, but not identical, to its use in government services such as the railways: in TISCO 'covenanted' referred not to a particular officer grade but to the terms of employment, thus covering both technical experts (officers) and skilled workers (operators). The term was not used for Indians.[29] However, like their counterparts in the government services, TISCO's covenanted employees were paid high salaries (in particular, they earned more than the few Indians who reached the same positions, as shown later in this chapter) and granted other benefits such as company housing. Bachelors were housed together in company-built bungalows or lodged at a local hotel, while those who were married had independent houses.[30] A race course was set up for the entertainment especially of the imported staff, whose sporting adventures were made possible by their high salaries.[31]

Like the construction engineers before them, the covenanted employees were drawn from several countries. During his extensive travels many years earlier, Jamsetji Tata had formed views on the strengths of various nations in particular aspects of steel-making. He had communicated these views to his son, suggesting that workers for the plant's departments be sourced according to these strengths.[32] Visiting the works in 1911, the journalist Lovat Fraser reported a division of labour that followed this advice:

[28] This was in addition to several thousand unskilled labourers recruited locally. Harris [with Fraser], *Jamsetji Nusserwanji Tata*, p. 202; Lala, *Romance of Tata Steel*, p. 27.

[29] A footnote at the start of the next section demonstrates in detail that 'covenanted' referred only to foreigners. The usage of 'covenanted' in the railways to mean the officer grade is mentioned in Chapter 4.

[30] Keenan, *Steel Man*, p. 38; Lillian Ashby (with Roger Whately), *My India: Recollections of Fifty Years* (Boston: Little, Brown and Company, 1937), pp. 359–60. Lillian Ashby's husband Robert was a high-ranking police officer in Jamshedpur in the interwar period.

[31] Ashby, *My India*, p. 298; Keenan, *Steel Man*, chapter 3.

[32] Keenan, *Steel Man*, p. 38; also mentioned in Headrick, *Tentacles of Progress*, p. 371, citing Keenan.

Mr [R. G.] Wells [of the USA] was General Manager, and his chief assistants in the management, as well as the Blast Furnace Superintendent and
his staff, were all Americans. The crew of the steel works [open hearth
furnaces] and their superintendent were Germans. The superintendent
and crew of the rolling-mills were English. The clerical staff was chiefly
composed of Bengalis and Parsees, and there were a few extremely efficient Parsees in the various mechanical departments. There were a certain
number of Austrians, Italians, and Swiss, while Chinese were working as
carpenters and in the pattern-shops.[33]

It appears that the superintendents of each department brought along
workers from their respective countries to form their crews. Thus, the
blast furnace department had a group of steel operators from eastern
Pennsylvania, probably schooled in the mills of Pittsburgh, while the
German head of the open hearth department (according to the possibly biased testimony of an American superintendent) even selected
for an important position 'a man who had had no steel-making experience at all; he had, however, been an officer in the Kaiser's own
regiment'.[34]

As was already evident to Lovat Fraser in 1911, the plant at its
highest levels was managed almost exclusively by American experts
schooled in the steel industry of their home country. In its first three
decades, TISCO's General Managers were engineers or steel experts
from the USA. The General Manager was the most important func-
tionary in Jamshedpur, having charge of the entire operation of the
works and the township. As an experienced construction engineer,
R. G. Wells, the first General Manager, was a suitable choice for the
early years when the works were being built and inaugurated. His
successor, T. W. Tutwiler (General Manager 1916–25), was a veteran
of the steel works in Gary, Indiana.[35] John L. Keenan (General Manager 1930–7) was also a former Gary employee. He had worked there
under Tutwiler, who recruited him to TISCO in 1913 as a foreman in
the blast furnace department. The Irish-American Keenan was born

[33] Harris [with Fraser], *Jamsetji Nusserwanji Tata*, p. 202.

[34] Keenan, *Steel Man*, pp. 38–40. The quote is from p. 40.

[35] Lala, *Romance of Tata Steel*, p. 65; for years of General Managership, see entry on
Tutwiler in 'Vanguards', available at http://www.tatasteel100.com/people/vanguards.
asp (accessed 1 July 2012).

in Roxbury, Massachusetts, in 1889. He studied mainly classics and economics at Boston Latin School and Yale University before he was trained in steel-making on the job, at the Wharton Steel Company of New Jersey and later at the Gary Steel Company. Keenan rose steadily in TISCO, going through various departments: he became in succession Assistant Superintendent (open hearth furnace), Assistant Superintendent (duplex plant), Superintendent (blast furnace), and General Superintendent of the works (in 1926, four years before he became General Manager).[36]

These high-level executives established a brisk, no-nonsense culture in the works. Tutwiler was known for his brusque demeanour. He is reported to have said to the Viceroy Lord Chelmsford (when asked politely if he was the General Manager), 'You're Goddam right'— an answer which nearly precipitated a diplomatic crisis.[37] Tutwiler believed that 'a steel works was no place for weaklings ... stern discipline, punctuality and hard work were essential to success ... the right to hire and fire and lay down the law was a God given right [of the managers]'.[38] Keenan comes across in his memoir as a tough-talking but fair-minded man who preferred a down-to-earth attitude to ostentation. He disapproved of the behaviour of a group of college-educated construction engineers from America who camped in Jamshedpur when the works were being extended. He noted that they

> didn't mix with the American and British steel operators. The old-timers ... danced the waltz, two-step and even the Lancers, the newcomers danced the fox-trot, the Lame Duck, and the tango.... They disdained our whiskeys and sodas, insisting on cocktails and other poisonous concoctions. The old crowd talked of horses and men; the youngsters of golf, dancing and women; they seemed to regard their stay in India as part of a world pleasure tour.[39]

[36] 'John L. Keenan, 54, Steel Authority', obituary in *New York Times*, 8 January 1944, p. 13; 'The Late Mr J. L. Keenan', obituary appearing in *TISCO Review*, March 1944, pp. 20–1; Keenan, *Steel Man*, pp. 1–3. Keenan was an astute observer, and his 1943 memoir, *A Steel Man in India* (cited earlier), is an important source on the environment in the TISCO works in the interwar years.

[37] Lala, *Romance of Tata Steel*, p. 65.

[38] J. R. D. Tata, a later Chairman of Tata Sons, commenting in 1956 on Tutwiler's approach, as quoted in *Industrial Development and Technology Absorption*, edited by Oba and Panda, p. 70.

[39] Keenan, *Steel Man*, p. 82.

The experts at the lower managerial levels came from Europe as well as America, and had varied educational backgrounds: some were university-educated, but many were practically trained.[40] An example is F. K. Bennett (1860–1932), who was born in Sheffield and went to the USA at the age of ten. Starting as a water boy, he was trained as a roller in the Pennsylvania Steel Company, eventually becoming Superintendent of Rolling Mills there before joining TISCO in 1914. An expert roller, he was Mills Consultant to the company for five years until his retirement in 1928.[41] European experts included Ernest Blaser, a Swiss engineer in charge of the boiler plants,[42] and E. R. Nicholson of Northumberland, who joined the company in 1918 as Master Pattern-Maker, later also becoming Assistant Foundry Superintendent.[43] R. M. Prowse, who was briefly Electrical Engineer at TISCO in the late 1930s, had a background spanning three continents. He was born in Devon, educated in South Africa, and further trained in England and the USA before joining the Tatas.[44]

Foreign-Trained Indians

Although the plant relied heavily on foreign experts in the early years, there is evidence to show that this was due more to practical concerns than to any prejudice against Indian engineers. As early as 1909, Axel Sahlin, whose firm built the plant, had mooted the idea of sending Indians to Europe or America for training in steel manufacture. However, the Board of Directors had reservations about the costs involved. An alternative was suggested by Bezonji Dadabhoy, who was associated with an older Tata concern, the Empress Mills (textiles) of Nagpur. Bezonji recalled that at his company a few Englishmen had initially been in charge, and had successfully trained Indians to take over from them. He continued:

I do not think it would help your Company, though it may help India generally, to send out young Indians to the U. S. to study and work in the Steel

[40] '[T]he European department heads.... Many of them had little or no higher education. They had learned what they knew by doing it.' Keenan, *Steel Man*, p. 137.

[41] *TISCO Review*, December 1932, p. 23.

[42] Keenan, *Steel Man*, pp. 44–6.

[43] 'Mr E. R. Nicholson', *TISCO Review*, December 1933, p. 20.

[44] 'Obituary', *TISCO Review*, April 1940, p. 291.

Works there. I would suggest apprentices being taken up and trained under American Experts, who may be brought out for starting and working the Steel Plant at Kalimati [the name of the railway station near Sakchi].

It was the question of costs that ultimately decided the issue. R. G. Wells, the General Manager at Sakchi, reported that '[u]pon completion of our recent estimate of capital expenditure we came to the conclusion that the matter of sending Apprentices [abroad] ... should be entirely abandoned'. It might be a good idea in ideal circumstances, 'but the Company certainly has no money to spend for such purposes'.[45]

Nevertheless, a few years later, TISCO was able to recruit from the growing band of Indian engineers and metallurgists who had begun to travel abroad on their own initiative for training. These students had turned to foreign countries as the existing education system in India was weighted too far in favour of 'literary and philosophic studies to the neglect of those of a more practical character'. Some of them received technical scholarships instituted by the colonial government, by private societies, or by princely states, while others supported themselves by working alongside their studies.[46]

The USA was a particularly attractive destination. This appears to have been so because it was easier to get industrial apprenticeships (as a supplement to formal education in the universities) in America than in Europe. The Hindusthan Association of America—a body set up by Indians in New York to guide prospective Indian students in the USA—claimed that 'America offers the best of opportunities to foreign students'. Foreigners were free to join American universities as well as the 'annual apprenticeship courses' in American factories. Such practical training was an essential part of a 'scientific and

---

[45] Various letters in GM's correspondence, April 1909. Participants include General Manager (R. G. Wells), Agents (Tata Sons), and Bezonji Dadabhoy. See sheets 13–26 in Box: General Manager's Correspondence, 1909, TSA. On Bezonji Dadabhoy (Mehta), see 'Sir Bezonji Mehta (1840–1927)', available at http://www.tatacentralarchives.com/history/biographies/02%20bezonjimehta.htm (accessed 1 July 2012).

[46] Indian Industrial Commission, 1916–18, *Report* (Calcutta: Superintendent Government Printing, 1918), chapter X, 'Industrial and Technical Education' (quoted text from p. 104); Bagal, *Pramatha Nath Bose*, pp. 91–2; Ross Bassett, 'MIT-Trained Swadeshis: MIT and Indian Nationalism, 1880–1947', *Osiris*, vol. 24 (2009): 212–30. The first of these sources is hereinafter cited as IIC *Report*.

industrial education', and it was only in America that foreign students could obtain such training 'without [being charged] any compensation or premium whatsoever'.[47]

Many such USA-trained Indian experts joined TISCO starting in the 1910s (in addition to some Indians who had been trained in Germany or England). A. C. Bose, who had graduated from the Carnegie Institute of Technology in Pittsburgh, joined the company as a chemical engineer, and eventually became chief chemist in place of an American. D. C. Gupta, a Bachelor of Science from Harvard, joined the open hearth steel department as third furnace hand around the time of the Great War. He later transferred to the coke oven department, and in two years became its Superintendent. After eight further years in the department, during which the Welsh foremen gave way to Indians, Gupta left to take up the post of Director of Industries for Bihar and Orissa (which he held from 1926 to 1933).[48]

An incident involving Gupta illustrates the importance placed by the management on employees' physical toughness. In his days as an operator in the open hearth department, Gupta earned Tutwiler's unspoken admiration for an act of physical bravado. On being 'insulted by a big Yorkshire foreman' in his days as an operator in the open hearth department, Gupta had

> said calmly to the man that he had been too long in America to take reflections on his parentage without a fight. The foreman made a pass at him. He dodged and landed a haymaker on the foreman's jaw, putting him out for the count.[49]

In contrast, John Keenan recalled the case of a Bengali, a talented mathematician, whom TISCO had intially rejected 'because he looked ill'. But the company hired him as a coke oven researcher after he had obtained a degree and two years' industrial experience in Germany. In

[47] The Hindusthan Association of America [New York City], *Education in the United States of America: For the Guidance of the Prospective Students from India to the United States, Bulletin No. 1, 2nd and revised edn* (New York City: n.p., 1920), pp. 2–3. British Library Shelfmark: General Reference Collection 8385.a.14.

[48] Keenan, *Steel Man*, pp. 134–5; entry on D. C. Gupta in Record of Services, *India Office List* for 1940.

[49] Keenan, *Steel Man*, p. 134.

a few months the new recruit was taken seriously ill and died, apparently unable to take the physical strain. Keenan felt that the management had recruited the unfortunate man 'against their own better judgement'.[50]

In one case, that of J. J. (later Sir Jehangir) Ghandy, the company itself appears to have arranged or supported an Indian's studies abroad.[51] Ghandy (1896–1972) was born and educated in Bombay, completing his BA Honours (physics and chemistry) at St Xavier's College in 1916 and his BSc Honours in chemistry at Wilson's College the following year. He then underwent a spell of practical training at the TISCO works in Jamshedpur before proceeding to the USA, where he studied business administration at Columbia University and metallurgical and steel works engineering at the Carnegie Institute of Technology (Pittsburgh). He returned to India and TISCO in 1921, this time as Metallurgical Engineer in the mill departments. He became the first Indian General Manager in 1938 and later one of the most important TISCO executives in post-Independence India.[52]

Some foreign-trained Indians were recruited in later decades too. C. S. N. Raju, who joined the company as Assistant Power Engineer in 1934, was a graduate of Madras University and an MS in Mechanical Engineering from the Massachusetts Institute of Technology. He had worked in America for two years and had been Inspector of Steam Boilers, Madras Government, for four years before joining TISCO.[53] S. K. Nanavati (c. 1907–1986), who joined the company in 1932, was British-trained. Nanavati did his BSc at the Royal Institute of Science, Bombay, then took the degrees of BMet (Hons) and MMet at Sheffield, followed by that of Doctor Ingenieur (Metallurgiste) from the University of Brussels in 1932. Nanavati went on to become TISCO's General Manager (1961) and its first Managing Director (1970).[54]

[50] Keenan, *Steel Man*, p. 137.

[51] Both Lala (*Romance of Steel*, p. 69) and Headrick (*Tentacles of Progress*, p. 373) write that Ghandy was 'sent' to America, whereas the other source on him, an obituary appearing in TISCO's newsletter (cited later), is ambiguous on whether Ghandy went to the USA on his own or was sponsored by the company.

[52] 'Sir Jehangir Passes Away: End of an Era', *TISCO News*, May 1972, pp. 2, 8, 14–17; Lala, *Romance of Tata Steel*, pp. 68–9.

[53] *TISCO Review*, June 1935, p. 488.

[54] 'Veteran Steelman [*sic*] Passes Away: Shavak Kaikhushru Nanavati', *TISCO News*, July–August 1986, p. 23.

The company's management took its time over applications from Indians studying or working abroad, but it also exhibited a degree of flexibility when it felt that a candidate was promising. This is well illustrated by the sequence of events leading up to the appointment of P. N. Mathur, who became one of the most prominent experts in the TISCO works in the 1930s. (Joining the company in 1927, he became Superintendent of the open hearth plant in 1931 and Superintendent of the duplex plant some months before his death from pneumonia in 1940.)

Born in Lahore, Prem Narain Mathur (1892–1940) had dropped out of medical college and made his way to the USA c. 1913. There he got himself into the Ford Motor Company at Detroit, completed a correspondence course in Metallography, and took practical classes in the YMCA night school. He was placed in Ford's research laboratory, and over the following years made a name for himself as an expert metallurgist.[55]

A little more than a decade after he had first arrived in America, Mathur decided to move back to India. He began corresponding with the Tatas in October 1924, writing first to Dorabji Tata and then several times to John Peterson, Director of TISCO, applying for a job. In March 1925, favourably impressed by Mathur's credentials, Peterson suggested to T. W. Tutwiler (the outgoing General Manager, who was returning to America) that he should meet Mathur in America and report on him. Meanwhile, Mathur was asked to specify his terms, and to furnish details of his personal background. He insisted that pay was not an important criterion, but when pressed, quoted a figure of Rs 1,500 per month. Tutwiler gave Mathur a glowing report after meeting him in September 1925 (he particularly approved of the fact that Mathur seemed 'like he would be willing to take off his coat [and] jump in'). However, C. A. Alexander, the new General Manager, felt that the company might not be able to afford Mathur's asking salary. Thereafter the matter appears to have stalled, despite a further query from Mathur. In January the following year Mathur, now working as Assistant Superintendent in the open hearth department at Ford,

<hr/>

[55] Capt. B. Dayal, 'Prem Narain Mathur (An Appreciation)', *TISCO Review*, May 1940, pp. 390–2. That Mathur enjoyed a strong reputation as a metallurgist is apparent from the many character references he furnished to TISCO. See the correspondence in Mathur's papers, cited later.

wrote to Peterson again. 'What I stand in need of,' he wrote, 'is a chance in India. If shown this favour I would be able to work out my destiny as well in India as I did in America.' Finally, some months later, the Board of Directors empowered the General Manger to offer Mathur up to Rs 1,750 per month. In due course he was appointed at Rs 1,500, and began work at Jamshedpur in late 1927.[56]

THE JAMSHEDPUR TECHNICAL INSTITUTE: INDIANIZING THE TISCO WORKS

Although Indians were already being recruited in the 1910s, TISCO soon felt it necessary to systematize the process. During and especially after World War I, the company decided to take into its own hands the training of Indians for supervisory and managerial positions. A number of factors made this a priority.

First, there was a pressing need to reduce costs. The covenanted (that is, foreign)[57] employees in the works were paid hefty salaries and production-based bonuses. Table 5.1 shows that their average annual pay (including bonuses) was many times that of the Indian

---

[56] Several letters and a note 'For Favour of Minutes' (date 13 December 1927, title 'Summary—Appointment of Mr Prem Narain Mathur'), Folder: 'P. N. Mathur, Appointment at Jamshedpur', Box No. 69: 'Private Papers: P. N. Mathur', TSA. The quotes are from the following letters: Tutwiler to Peterson, 20 September 1925; and P. N. Mathur to John Peterson, 20 January 1927.

[57] It is apparent from the use of the term in most sources that only foreign experts were referred to as 'covenanted' employees. Two instances will serve to confirm this:

1. A list of employees presented by the company to the Indian Tariff Board, which shows zero covenanted employees in the coke oven department for 1923–4 and 1924–5, around which time, according to John Keenan, that department was 'completely Indianized'. In other words, the Indians who replaced the foreign workers on the coke ovens, despite doing the same work, were not classified as covenanted. Enclosure VI, Indian Tariff Board, *Evidence Recorded During Enquiry Regarding the Grant of Supplementary Protection to the Steel Industry* (Calcutta: Government of India Central Publication Branch, 1925 [hereinafter cited as ITB 1925]), p. 66; Keenan, *Steel Man*, p. 134.

2. The point is reinforced by the following quote in a volume brought out by the company in 1958: 'The Works have been almost completely Indianized and today there are less than half a dozen members of the covenanted staff, who have made themselves at home in India and endeared themselves to its people.' Elwin, *The Story of Tata Steel*, p. 67.

uncovenanted staff. It must be noted that this is not a like-for-like comparison (the uncovenanted category included several levels of staff from skilled to unskilled labour, whereas the covenanted staff comprised members of the operating crews, supervisors, and managers). Nevertheless, the comparison shows that covenanted staff were an expensive resource. Further, their numbers dipped between 1912–13 and 1921–2, even as their average income quadrupled, indicating that while foreign workers could be replaced, the expensive managers and experts were still indispensable.[58] There were other overheads. The foreigners were also provided with benefits such as housing and passages from and to their home country;[59] and replacements had to be found for the German open hearth team, whose members were interned in Ahmednagar as enemy aliens when World War I broke out. Furthermore, the company was supplying the Government in excess of 20,000 tons of steel rails per year (at reduced prices) as railroads were built in the battlefields of Mesopotamia, and efficient production was the order of the day.[60] It was apparent that savings could be made if it were possible to find Indians who could attain the higher positions currently occupied by covenanted staff; when an Indian did so, he would be paid a maximum of two-thirds the salary drawn by a foreigner at the same level. This distinction was of a similar order to that between Britain- and India-recruited engineers in the PWD and railways (for example, see Chapter 3).[61]

[58] It should be noted that the numbers of covenanted workers rose again in the 1920s, but this should be seen against the massive expansion that the plant was then undergoing. According to Daniel Headrick, the highest number of foreigners was reached in 1924. Headrick, *Tentacles of Progress*, p. 372.

[59] Headrick, *Tentacles of Progress*, p. 371.

[60] Ray, *Industrialization in India*, p. 83; Ashby, *My India*, p. 299; Keenan, *Steel Man*, p. 14 and p. 45; Copy of letter from T. H. Holland to T. W. Holderness (Under-Secretary of State for India), dated 21 August 1918, in Indian Tariff Board, *Evidence Recorded During Enquiry into the Steel Industry*, vol. I: *The Tata Iron and Steel Company* (Calcutta: Superintendent Government Printing, India: 1924), pp. 96–7. The last of these sources is hereinafter cited as ITB 1924.

[61] Evidence of J. C. K. Peterson, T. W. Tutwiler and R. D. Tata at Jamshedpur, 18 August 1923, to the Indian Tariff Board, ITB 1924, pp. 275ff, here p. 280. The similarity with government service was noted by the President [of the ITB] during this interview (Evidence of J. C. K. Peterson, T. W. Tutwiler and R. D. Tata at Jamshedpur , ITB 1924, p. 280).

Table 5.1 Covenanted and uncovenanted employees of TISCO in various years
Number and average pay (aggregate of coke ovens, blast furnaces, open hearth, blooming mill, 28" mill and bar mills)

Year	Covenanted employees	Total wages and bonus (Rs)	Average annual pay (covenanted) (Rs)	Uncovenanted employees	Total wages (Rs)	Average annual pay (uncovenanted) (Rs)
1912–13	140	459,714	3,284	3,917	863,144	220
1915–16	75	637,784	8,498	4,243	1,120,284	264
1921–22	74	964,592	13,035	9,924	2,979,948	300

Source: Calculated from tables in Indian Tariff Board, *Evidence Recorded During Enquiry into the Steel Industry*, vol. I: *The Tata Iron and Steel Company* (Calcutta: Superintendent Government Printing, India, 1924), Statement No. I, pp. 109–11. For rates of bonus, see Indian Tariff Board, *The Tata Iron and Steel Company*, pp. 198–200.

Note: The figures for number of employees in this table are not strictly comparable with those for the following decade, by which time the extensions to the plant had become operational.

The second factor that encouraged the creation of training facilities for Indians was related to TISCO's post-war plea for economic protection. While the colonial government granted the company's request for interwar protection from cheap Belgian and German steel imports by raising protective tariffs,[62] in return, TISCO had to submit to periodic scrutiny by the Indian Tariff Board (ITB). The ITB had to be satisfied that the company was doing all it could to justify the burden on the Indian consumer, who would pay higher prices for steel, and the Indian taxpayer, whose money would be used for bounties granted to TISCO in this period. The company was expected to demonstrate that it was keeping its costs down, and the Tariff Board saw the replacement of expensive foreign employees by Indians as one of the ways to do this.[63] Further, in the opinion of one ITB official, 'There [was] no question in which the tax-payer [was] more keen than the scope for employment of Indians.'[64]

Third, the company had inaugurated a major expansion project in 1916, following upon the government's high consumption of TISCO steel and the establishment of other steel-consuming industries during the war.[65] As we saw earlier, the Greater Extensions aimed to increase the existing output of the works by a factor of five, and over the next few years it commissioned new plants that used improved production techniques such as the duplex process.[66] Although more foreign workers were recruited to fill the new crews,[67] importing experts was an expensive option. Consequently, the company needed as many qualified Indians as it could get.

Fourth, university courses in metallurgy, mechanical engineering, and electrical engineering were still rare in India. The major engineering colleges at Roorkee, Madras, Sibpur, and Poona were primarily

[62] T. Roy, *The Economic History of India*, p. 235.

[63] Evidence of Peterson, Tutwiler and R.D. Tata, ITB 1924, p. 280.

[64] See the exchange between Dr Matthai (ITB) and Mr Peterson (Tata Sons) in ITB 1925, pp. 102–3. The quote is from Dr Matthai on p. 103.

[65] Ray, *Industrialization in India*, p. 87.

[66] Lala, *Romance of Tata Steel*, pp. 37–9; Keenan, *Steel Man*, p. 68; *First Report of the Indian Tariff Board Regarding the Grant of Protection to the Steel Industry* (place and publisher illegible: [1924]), p. 7. (Accessed via Digital Library of India, www.dli.ernet. in, on 2 January 2012.)

[67] Headrick, *Tentacles of Progress*, p. 372.

intended to produce civil engineers for government service.[68] In the interwar period courses in metallurgy and mechanical engineering began to be offered at privately run institutions such as the Banaras Hindu University and the Bengal Technical Institute, but these were few and far between.[69]

All these factors favoured the establishment of new, in-house training facilities for Indians. They also dictated the form that such training should take: a combination of university-style theoretical learning with practical apprenticeship in the works. As early as 1916, the Tatas initiated an idea for a metallurgical school in Sakchi, offering a course of two years' theoretical instruction and two years' industrial apprenticeship. According to the proposal, the Bihar and Orissa Government would set up the school, the TISCO works would be used as a site for practical training, and students who passed out would add to the industrial manpower of the province, some of them obtaining jobs in TISCO itself.[70] The plan was placed before the Indian Industrial Commission and met with its 'general approval',[71] but, for reasons that are not clear, did not come to fruition. However, it provided the template for a plan that did materialize.

This was the Jamshedpur Technical Institute (JTI). Set up in 1921 under the primary control of the Tatas, it comprised 'laboratories, lecture halls, classrooms and a library'.[72] It was not meant to be a 'trade' or 'industrial' school, the sort that typically admitted boys from an artisanal background and taught them a practical skill like blacksmithy or carpentry, or other semi-skilled factory work.[73] Instead, it

[68] Arun Kumar, 'Colonial Requirements and Engineering Education: The Public Works Department, 1847–1947' in *Technology and the Raj: Western Technology and Technical Transfers to India 1700–1947*, edited by Roy MacLeod and Deepak Kumar (New Delhi, Thousand Oaks, and London: Sage Publications, 1995), pp. 216–32.

[69] Table 5.5 (later in this chapter) has more on the courses offered at these two institutions.

[70] Evidence of F. Walford, Witness No. 40, in Indian Industrial Commission (IIC), *Minutes of Evidence 1916–17*, vol. I: *Delhi, United Provinces and Bihar and Orissa* (Calcutta: Superintendent Government Printing, 1917), pp. 316–18.

[71] IIC *Report*, p. 133, paragraph 172. It must be noted here that one of the Commission's members was Dorabji Tata, Chairman of Tata Sons (IIC *Report*, p. xvi), although it is not known if he participated in this particular decision.

[72] Keenan, *Steel Man*, p. 135.

[73] See Evidence of F. Walford in IIC, *Minutes of Evidence*, p. 315, on the nature of 'industrial schools'.

was designed to produce employees rich in theoretical knowledge as well as practical experience, who would be fit to take on supervisory and managerial roles in the TISCO plant. The JTI's existence in the interwar period may be viewed as having two distinct phases falling on either side of the year 1935. I will discuss each in turn.

The JTI's First Phase: The Three-Year Programme

In the first phase of its existence, 1921–34, the JTI ran a three-year training course for students of university-going age. The structure of the course and the selection of staff and students were such as to enable a combination of scientific education and hands-on work. The instructors were university-trained men, including two (presumably British) Assistants, both Bachelors of Metallurgy from Sheffield, and an Indian BSc (Calcutta). The Director of the Institute, W. Saunders, also had a Sheffield connection: he had been an apprentice at Vickers Ltd in that city, in addition to holding a BSc from London.[74] It is not clear whether the search for senior staff included the USA and other countries, but correspondence of the time confirms that advertisements were placed in the *Times* in Britain and 'private enquiries' made at British universities.[75] Some years later, the British instructors were joined by Indians who played important roles in the running of the JTI. S. N. Roy, one of the students of the JTI's 1922 batch, joined the staff of the Institute briefly upon graduation.[76] Continuing the Sheffield motif, he was then sent with a scholarship from Dorabji Tata to Sheffield University, where he earned a BMet degree before returning in 1928 to take up the post of Instructor at the JTI.[77] Roy

[74] Statement no. IX, ITB 1924, p. 121.

[75] [H. Treble] to S. M. Marshall, TISCO, 24 March 1921, Folder: T1/A1, 1921 to 1960, Box: Technical Training, T1/A1 (1921–60), A1 (1962–4, 1965–8, 1966–84), TSA.

[76] Statement 108, enclosure (2), *The Applications Received by the Indian Tariff Board from the Tata Iron and Steel Company, Limited, and Other Companies in Connection with the Statutory Enquiry Regarding the Grant or Continuance of Protection to the Steel Industry in India after the 31st March 1927, Together with the Questionnaires Issued by the Board and Other Relevant Papers* (Calcutta: Government of India Central Publication Branch, 1926), p. 160. This source is hereinafter cited as ITB 1926.

[77] W. Saunders, JTI Director, to the TISCO General Superintendent, 26 October 1927; S. N. Roy (in Sheffield) to JTI Director, 13 November 1927. Both letters in Folder: T1/A1, 1921 to 1960, Box: Technical Training, T1/A1 (1921–60), A1 (1962–4, 1965–8, 1966–84), TSA.

later went on to become Superintendent of Training (that is, head of the JTI).[78] Another Indian who took a keen interest in the Institute was P. N. Mathur, who in his days at Ford had conducted classes in metallography for his colleagues. Mathur served as President of the Technical School Committee.[79]

As for students, the JTI invited applications from school-leavers with a science background: candidates had to have passed the Intermediate Science (ISc) examination (administered at the end of a two-year course between school and university levels). These students would naturally be literate in English and of university standard. No prior factory experience was required, as the JTI's course itself provided for in-depth practical learning.

In the first two years of the three-year course of instruction, students alternated between the Institute (where they were instructed in theoretical subjects) one week and the TISCO works the next (first as observers, then as apprentices). In the third year, having gained a firm theoretical knowledge base, they were apprenticed full-time to the works. During their practical training students reported to the Works Superintendent, while their overall responsibility was to the Institute. After the second year, students decided to specialize in a particular branch of steel-making: coke ovens, blast furnace, open hearth, or sometimes rail mills.[80] Unlike the gentleman-generalist engineer of the PWD, the steel expert in the making was expected, from this point on, to concentrate all his energies on his chosen specialism. He had to familiarize himself with 'all the technical literature ever published in England, Germany, France and the United States, so that he becomes a sort of walking encyclopedia on [his chosen subject]', able to answer any question put to him by the superintendent to whom he was apprenticed in the works. In time it was necessary for the student to be able to recall relevant information at a moment's notice, for

[78] See S. N. Roy's designation in several memos/letters sent by him in 1937 and 1938 in B. K. Mukherjee File, Folder: 1937, Box: Technical Training: A & B Class Apprentices, 1937, '38 and '39 batches, TSA.

[79] Capt. Dayal, 'Prem Narain Mathur (An Appreciation)'; Keenan, *Steel Man*, p. 139.

[80] Evidence of Saunders, JTI Director, in ITB 1924, pp. 302–5; Keenan, *Steel Man*, pp. 135–6.

'[b]y the time he [had] a second thought in an emergency, he may be too dead to need it'.[81]

Like the other steel experts, the students of the JTI had to be in excellent physical condition: the ability to spend long hours in the works in high temperatures and potentially dangerous situations was as important for potential officers as it was for the plant's workers. To this end a medical examination was an important part of the selection process for entry into the Institute. Further, in promoting students from one year to the next, special consideration was given to students who were physically tough and industrious in the plant even if they were below par in their theoretical studies.[82] Students were also encouraged to play sports in their free time.[83]

The JTI was not a profit-making organization but one designed to train employees for the TISCO works, moulding them according to the company's needs. The students were charged no fees, received a stipend of Rs 60 a month, and were given subsidized lodging. The company provided them with the necessary books, which they paid for later if they were selected for a job in the works upon graduation.[84]

Although it was run primarily by the Tatas, the Institute received financial support from some provincial governments during the first decade of its existence (discussed later). Table 5.2 shows details of the costs of running the JTI, and external contributions, in the first two years of operation (1921–2 and 1922–3). The governments of Bihar and Orissa (Rs 25,000 per annum) and Bengal (Rs 10,000 per annum) were the major contributors. The princely state of Mysore sent a few students, paying Rs 2,000 per student; the Calcutta firm of Bird & Co. likewise paid for one student's training. (They were investing in personnel for their own use: sponsored students returned to work in Mysore state and Bird respectively after their training.[85]) Officials of the provincial governments also participated in the first stage of selection. Saunders, the Director of the JTI, reported that the

[81] Keenan, *Steel Man*, p. 136.

[82] See Evidence of Saunders, ITB 1924, p. 303.

[83] Keenan, *Steel Man*, p. 135.

[84] Evidence of Saunders, ITB 1924, p. 304.

[85] For examples of students transferred to Mysore Distilleries and Bird & Co. upon graduating from the JTI, see Statement no. 108, enclosures (1) and (2), ITB 1926, pp. 159–60.

Table 5.2 Expenditure on and contributions to the Jamshedpur Technical Institute (JTI) in its first two years

	1921–2 (Rs – Annas – Paise)	1922–3 (Rs – Annas – Paise)
Recurring Expenditure	42,710 – 5 – 0	71,009 – 3 – 3
Contributions		
Mysore	4,800 – 0 – 0	4,800 – 0 – 0
Bihar and Orissa	8,333 – 5 – 4	25,000 – 0 – 0
Bengal	-	10,000
Bird & Co.	-	1,375 – 0 – 0*
Sir Ratan Tata Trust	-	15,000 – 0 – 0
Remainder, borne by Steel Company	29,576 – 5 – 8	14,834 – 3 – 3
Actual Capital Expenditure	130,519 – 4 – 8	11,277 – 0 – 8
Receipts from Bihar and Orissa	100,000 – 0 – 0	-
Remainder, borne by Steel Company	30,519 – 4 – 8	11,277 – 0 – 8

Source: Indian Tariff Board, *Evidence Recorded during Enquiry into the Steel Industry*, vol. I: *The Tata Iron and Steel Company* (Calcutta: Superintendent Government Printing, India, 1924), Statement no. IX, p. 121.

*This figure is marked 'No stipend' in the source. This probably means that the amount paid was sufficient to cover the expenses of the student sent by Bird & Co., but did not include a stipend for the student.

Directors of Industries in Punjab and Madras examined candidates from their respective provinces and sent the best qualified ones on to Jamshedpur, where they came before a Selection Committee for a further round of elimination and a medical examination.[86]

The provinces' decision to contribute monetarily to the Institute may be understood in the light of the constitutional reforms of 1919, which had introduced the system of dyarchy, and made Industries a 'transferred' subject under provincial ministers.[87] Possibly, the provincial governments wished to promote the JTI as an opportunity for industrial training for their candidates (some seats were reserved for candidates from donor provinces). Certainly in the case of Bihar and

[86] Evidence of Saunders, p. 303.

[87] A. Z. M. Iftikhar-ul-Awwal, *The Industrial Development of Bengal: 1900–1939* (Delhi: Vikas Publishing House, 1982), p. 59.

Orissa, technical colleges of the province sent their students to the JTI and TISCO to complete the apprenticeship component of their course. These students were also paid stipends and in some cases employed in the works at the end of the apprenticeship, but their programme of training was distinct from the regular three-year course of the JTI.[88] It will be recalled that the Bihar government had itself originally planned to establish a new institute, the students of which would undergo practical training at TISCO. It is plausible that by contributing monetarily to the JTI, they were seeking to achieve a similar objective at a lower cost.

From the start the JTI (and the prospect of TISCO jobs afterwards) proved immensely popular. The aim was to select around twenty-five students each year,[89] but the demand for places was many times greater. When applications were invited for entry in the second year of the JTI's operation, more than 2,600 candidates applied from numerous provinces and princely states across India (see Table 5.3).

The Institute was extremely selective. Table 5.3 shows that only 172 candidates of the 2,638 who applied for admission in 1922 were shortlisted to appear for an interview at Jamshedpur. Following the interview, only twenty-nine were finally admitted to the JTI. The programme was challenging, and several students could not keep up with it and had to leave the Institute.[90] Of the students admitted in the 1921 and 1922 batches, only 50 per cent graduated successfully.[91]

The students of the JTI, as we have seen, came from all over the country. Table 5.4 shows that in aggregate terms, the largest number came from the provinces of Bihar and Orissa and Bengal. These were in a sense the 'home' provinces of TISCO (Jamshedpur was located in Bihar and not far from Bengal); and, as we have seen, they both provided grants to the Institute (Table 5.2) and had places reserved for their candidates. In examining the JTI's figures, Professor V. G. Kale and Sir P. P. Ginwala of the Indian Tariff Board expressed surprise at

---

[88] Files of various students, Box: Technical Training: Bihar & Orissa Apprentices, 1927–33 Batches, TSA.

[89] Evidence of Saunders, ITB 1924, p. 303.

[90] Evidence of Saunders, ITB 1924, p. 303.

[91] Statement no. 108, 'Jamshedpur Technical Institute: Notes on Institute Results', ITB 1926, p. 158.

Table 5.3 Applicants to the JTI for November 1922, by province/princely state

Province	No. of application letters	No. of students with good physique and otherwise qualified [that is, ISc diploma and approved by province]
Bihar and Orissa	224	23
Bengal	885	64
Central Provinces	57	-
Madras	641	29
Bombay	63	4
United Provinces	173	7
Assam	43	6
Punjab	380	31
Central India	11	1
NW Province	13	-
Berar	1	-
Mysore	53	1
Burma	2	-
NW Frontier Province	11	1
Coorg	8	1
Baroda	4	-
Deccan	9	-
Cochin	14	2
Indore	3	-
Travancore	8	2
Cooch Bihar	2	-
Bikaner	1	-
Ajmer	8	-
Sindh	4	-
Kashmir	14	-
Jodhpur	3	-
Rajputana	3	-
Total	2,638	172

Source: Indian Tariff Board, *Evidence Recorded During Enquiry into the Steel Industry*, vol. I: *The Tata Iron and Steel Company* (Calcutta: Superintendent Government Printing, India, 1924), p. 123.

the low share of students from Bombay (with a reputation as an industrial centre rivalling that of Calcutta).[92] Possibly, alternative industrial

[92] ITB 1924, pp. 304–5. For full listing of ITB members, see *First Report of the Indian Tariff Board*, p. 2.

Table 5.4   Province-wise origin of JTI students until 1926

| Province | Graduated (as of 1926) | | In the Institute (as of 1926) |
	1921 batch	1922 batch	Total (including 1923, 1924 and 1925 batches)
Bihar and Orissa	7	4	21
Bengal	2	4	14
Assam	–	1	1
United Provinces	1	–	2
Madras	1	2	7
Central Province	1	–	2
Punjab	1	4	8
Bombay	–	–	3
Total	13	15	58

Source: The applications received by the Indian Tariff Board from the Tata Iron and Steel Company, Limited, and other companies in connection with the Statutory Enquiry regarding the grant or continuance of protection to the Steel Industry in India after the 31st March 1927, together with the questionnaires issued by the Board and other relevant papers (Calcutta: Government of India Central Publication Branch, 1926), Statement no. 108, Enclosure (5): 'Provincial List', p. 163.

opportunities in Bombay (with its textile mills) made the Institute less attractive to its students. In any event, very few applications were received from that province (thirty-two for 1921 and sixty-three for 1922).[93] Ginwala and Kale's surprise underlines the fact that Bombay was the exception that proved the rule that JTI places were coveted. The company's statement furnished in response to their query underlines its high standards and the qualities it prized in applicants. Of the sixty-three applicants from Bombay for 1922, the statement said, only four were selected for the next round. Two of these went to Jamshedpur. One failed the medical examination, and the only one selected, T. R. Kapadia, went on vacation at the end of April 1923 and did not return. The statement concluded drily: 'It appears that he found the work too strenuous.'[94]

From the mid-twenties, graduates of the JTI were ready to join the various departments and work their way up to the 'the more expert and

[93] Statement no. IX, ITB 1924, pp. 121–2, here p. 122.
[94] Statement no. IX, ITB 1924, pp. 121–2.

responsible class of work'.[95] While they were not guaranteed employment in the Tata works, many of those who completed the course successfully were offered contracts. They began at a minimum of Rs 200 a month, and were contracted for five years in the first instance.[96]

Most of these graduates began as foremen or assistant foremen[97] and moved up into the managerial grades within a few years. In 1926 the company reported that some of the JTI graduates who were now in the works (since 1924 or 1925) had already been promoted;[98] by 1928, three JTI graduates were 'actually operating furnaces'.[99] The rate of ascent through the ranks varied by the individual. One who rose rapidly was S. Sambasivan, who was appointed Chief Inspector in the Inspection Department upon graduation from the JTI.[100] In 1932 he was already Superintendent of the order department (the first JTI graduate to reach that rank).[101] In time others were promoted too. Keenan reported in 1943 that the current superintendents of the open hearth furnaces and the duplex plant were graduates of the first JTI batch.[102]

As early as 1926 the company reported to the Indian Tariff Board that 'the contracted men [that is, JTI graduates] are building up an encouraging reputation which undoubtedly justifies the system being adopted'.[103] The Institute kept up a steady supply of personnel to the works in the following years: as of 1932, the total number of contracts awarded stood at eighty-eight.[104] Furthermore, between two and four

[95] Statement no. 108, 'Jamshedpur Technical Institute: Notes on Institute Results', ITB 1926, p. 158.

[96] Evidence of Saunders and John Peterson, ITB 1924, pp. 302–5.

[97] Statement no. 108, enclosure (1), 'Statement Showing the Designation and Work of the Students Who Were Recruited in 1921 for Metallurgical Apprenticeship and Completed the Course', ITB 1926, p. 159.

[98] Statement no. 108, 'Jamshedpur Technical Institute: Notes on Institute Results', ITB 1926, p. 158.

[99] Keenan, *Steel Man*, p. 140.

[100] Statement no. 108, enclosure (1), 'Statement Showing the Designation and Work of the Students Who Were Recruited in 1921 for Metallurgical Apprenticeship and Completed the Course', ITB 1926, p. 159.

[101] *TISCO Review*, Dec 1932, p. 17.

[102] Keenan, *Steel Man*, p. 135.

[103] Statement No. 108, 'Notes on Institute Results', ITB 1926, p. 158.

[104] TISCO Annual Report for 1931–2 (Directors' Report), p. 8. (Consulted at TSA.)

of the graduates from most years were sent for a further spell of specialized training in particular branches of steel manufacture in plants in Germany and England.[105] .

John Keenan further endorsed the competence of the JTI graduates. In 1932, when he was General Manager, the management decided to set the JTI men a challenge. The effects of the Depression had begun to be felt and the pressure on production was reduced, so it was not critical to keep all the open hearth furnaces as well as the duplex plant running at full capacity. Some of the foreign technicians in the open hearth department were 'sent ... home' (presumably this means they were dismissed or did not have their contracts renewed), and others were transferred to the duplex plant. A group of JTI graduates was placed in charge of the open hearth furnaces under the direction of Prem Mathur. 'Within a year,' Keenan recalled, 'they were averaging 34,000 tons a month and had set a record for one month of 37,000 tons, or nearly double the American and British Tata record of 1919.'[106]

The JTI's Second Phase: The Graduate Trainee Programme

Despite the success that attended the JTI's early years, a few problems emerged. To begin with, the inability of many students to complete the three-year programme suggested that despite the strict admission procedure, not all students possessed the required aptitude. Saunders, the Institute's Director, observed in 1924: 'We find that the I.Sc. [Intermediate Science] Indian qualification is not a very high qualification.'[107] To add to this, steel manufacturing in the following years was seen as relying increasingly on scientific theory, which also suggested the need to admit better qualified students. As the Superintendent of Training put it in 1935, 'the days of rule-of-thumb methods in industry have gone by'.[108]

[105] See section on Technical Institute in Directors' Report, TISCO Annual Report, 1931–2 through 1933–4.

[106] Keenan, *Steel Man*, p. 140.

[107] Evidence of Saunders, ITB 1924, p. 303.

[108] 'Technical Institute at Jamshedpur: Superintendent's Speech at Annual Prize Distribution', *TISCO Review*, July 1935, pp. 561–2, here p. 561.

Further, the obligation to take in a certain number of students from donor provinces meant that the Institute's management did not have an entirely free hand in the selection of students. This, Keenan implies in his memoir, was another factor diluting the quality of admitted students.[109] Finally, the JTI sometimes had to take on apprentices from provincial technical schools outside of its regular three-year programme. An incident that occurred in 1931 indicates that the Institute found this arrangement one-sided, especially when its autonomy was challenged. In that year the Orissa School of Engineering in Cuttack sent four of its students to the TISCO works via the JTI for the start of a two-and-a-half-year apprenticeship. Soon after his first communication, Sohan Lal, the Principal of the Orissa School, wrote to amend the list, asking the company to accept a student named Mahadeb Naik in place of K. V. Gopala Rao.[110] While Lal insisted that Gopala Rao had been named '[b]y a mistake',[111] Saunders of the JTI suspected a different motive. Rao, he said in an internal memo, was 'the best of the bunch' sent from the Orissa School, and the Principal now wanted to replace him, 'obviously on second thoughts'. He suggested that Rao's roots in the presidency of Madras might be the cause of the changed decision.[112] During the ensuing negotiations, the company stressed its right to choose the better candidate (especially as he was the one sent originally), pointing out that it paid the apprentices a stipend for two and a half years (and might employ them at the end). 'In view of the facilities which this Company grants to the students from Ranchi Technical School and the Orissa School of Engineering, we do not think our request is unreasonable.'[113]

[109] Keenan, *Steel Man*, p. 138.

[110] Sohan Lal (Principal, Orissa School of Engineering) to the General Manager, TISCO, 12 January 1931, K. V. Gopala Rao File, Box: Technical Training: Bihar & Orissa Apprentices, 1927–33 Batches, TSA.

[111] Sohan Lal to General Manager, TISCO, 28 January 1931, K. V. Gopala Rao File, Box: Technical Training: Bihar & Orissa Apprentices, 1927–33 Batches, TSA.

[112] W. Saunders to General Superintendent [TISCO], 19 January 1931, K. V. Gopala Rao File, Box: Technical Training: Bihar & Orissa Apprentices, 1927–33 Batches, TSA.

[113] Signed by H. Chew for General Manager, TISCO, to Principal, Orissa School of Engineering, 30 January 1931, K. V. Gopala Rao File, Box: Technical Training: Bihar & Orissa Apprentices, 1927–33 Batches, TSA. TISCO eventually won the argument and the Orissa School allowed Gopala Rao to continue his apprenticeship with the company (Sohan Lal to General Manager, TISCO, 2 February 1931, in the same file).

In order to address these challenges, the JTI was reorganized substantially starting with the 1935 batch of trainees.[114] The Institute now dispensed with the grants from provincial governments, and 'was free to pick men solely on merit and physical stamina'.[115] The ISc was no longer a sufficient qualification to enter the Institute: the programme was redesigned as a two-year course for graduates.

The Institute was not limited any more to instruction in metallurgy, but also included electrical and mechanical engineering. According to the prospectus for the year 1936–7, students were classified as A1, A2, or B Class Apprentices according to their prior qualifications, as follows:

B: 'A Degree or Diploma in Mechanical or Electrical Engineering or Metallurgy of a recognised Indian or Foreign University, Technical Institute or College.'

A1: 'An HONOURS or FIRST CLASS Degree or Diploma in Mechanical or Electrical Engineering or Metallurgy of a recognised Indian or Foreign University, Technical Institute or College, preferably accompanied by Works experience ABROAD.'

A2 (the highest class): 'An HONOURS or FIRST CLASS Degree or Diploma in Mechanical or Electrical Engineering or Metallurgy of a recognised Indian or Foreign University, Technical Institute or College, accompanied by not less than 6 months continuous practical experience after graduation in an IRON & STEEL WORKS ABROAD.'

The age limit to apply to the JTI was twenty-seven years for graduates of foreign universities and twenty-four for those of Indian universities. A2 Class Apprentices could receive a stipend of up to Rs 200 a month at the discretion of the company. A1 and B Class Apprentices were paid Rs 75 per month and Rs 50 per month respectively. Apprentices also received subsidized housing and 'free medical attention'

[114] Superintendent of Training to Superintendent S.M.S.3, 22 March 1961, B. K. Mukherjee File, Folder: 1937 Batch, Box: Technical Training, A&B Class Apprentices, 1937, 1938, 1939 Batches, TSA. See also Directors' Report in Annual Report for 1934–5, p. 8 (consulted at TSA). This is the first Annual Report to mention A and B Class apprentices.

[115] Keenan, *Steel Man*, p. 138.

(under certain conditions). A clause stated that admission would not in any way depend upon a candidate's province of origin or residence.

While the educational qualifications of entrants to the JTI were now higher than in the Institute's first phase and the course of instruction shorter, the programme maintained several of its essential features. As before, the course was a mix of theoretical instruction and practical training in the works.[116] Students passed through a vast number of departments for short periods of time before specializing in the work of any one department, where they would spend six months undergoing training.[117] The emphasis on physical fitness was unaltered: 'Applicants must be of robust physique and be fit to perform hard manual labour.' A student could be asked to leave the Institute if he 'fail[ed] at any time to satisfy the Management in regard to the standard of [his] work, conduct, attendance or health'.[118] In the application forms of several students (archived in their respective files), the entries against 'Proficiency in College Athletics' are specially marked in pencil, probably by a member of the selection committee.[119]

The average number of A and B Class apprentices admitted in each batch between the years 1936 and 1939 (both inclusive) was thirteen.[120] The educational backgrounds of some of the early graduate trainees, and the jobs they obtained later, are shown in Table 5.5.

The table confirms the fact that courses in branches of engineering other than civil were still in their infancy in interwar India,[121] with

[116] TISCO Ltd, 'The Jamshedpur Technical Institute: Prospectus: Session 1936–1937', A. P. Mitter File, Folder: 1937 Batch, Box: Technical Training, A&B Class Apprentices, 1937, 1938, 1939 Batches, TSA. In addition to the graduate trainee course, the Tatas' apprentice programme, in which sons of employees were trained in trades such as fitting, welding and machining, was merged with the JTI. These students were now termed C Class Apprentices (Keenan, *Steel Man*, pp. 137–8).

[117] See, for instance, V. M. Subramanian file, Folder: 1937 Batch, Box: Technical Training, A&B Class Apprentices, 1937, 1938, 1939 Batches, TSA.

[118] TISCO Ltd, 'The Jamshedpur Technical Institute: Prospectus: Session 1936–1937', A. P. Mitter File, Folder: 1937 Batch, Box: Technical Training, A&B Class Apprentices, 1937, 1938, 1939 Batches, TSA.

[119] Files of students listed in Table 5.5. Sources are the same as those cited for Table 5.5.

[120] Calculated from figures in Directors' Reports appearing as part of the TISCO Annual Reports for the relevant years (TSA).

[121] See Kumar, 'Colonial Requirements and Engineering Education', p. 228.

Table 5.5 Profiles of some graduate trainees, JTI

Name and province of domicile	Year of joining JTI (usually January)	Class of Apprenticeship (if known)	Prior Qualifications*	Job held (if known)
A. P. Mitter (Bengal)	1937	B	Mechanical Engineering (four-year Diploma), Jadavpur	–
B. K. Mukherjee (Bihar)	1937	B	Electrical Engineering (Diploma), Jadavpur	As of 1961: First Converterman in New SMS.3 Bessemer Converter Section
V. M. Subramanian (Sind)	1937	A1	BSc in Mechanical and Electrical Engineering, BHU; G.I.E.E. (membership received while in the Institute)	Assistant Sales Engineer, Wheel Tyre & Axle Plant (as of 1942); Assistant Sales Manager, TISCO, Madras (as of 1946)
U. A. Prabhu (Madras)	1938	B	BSc in Metallurgy, BHU	Foreman, C1 Foundry (as of 1948)
N. K. Ganguly (Punjab)	1938	A1	Diploma of Maclagan Engineering College, Lahore (Electrical Engg.) and BSc in Engineering, Punjab University	–

(*Cont'd*)

Table 5.5   *(Cont'd)*

Name and province of domicile	Year of joining JTI (usually January)	Class of Apprenticeship (if known)	Prior Qualifications*	Job held (if known)
K. P. Mahalingam (Bombay)	1942	–	Diploma in Mechanical Engineering, VJTI	–
P. V. Ramachandran (Madras)	1942	A1	BSc in Mechanical and Electrical Engineering, BHU	Foreman Machine Shops I (as of 1946)

Sources: Files of the concerned students (including correspondence and applications for admission to the JTI), in

  i) Folders: 1937 Batch and 1938 Batch, Box: Technical Training, A&B Class Apprentices, 1937, 1938, 1939 Batches, Tata Steel Archives, Jamshedpur (TSA).

  ii) Folders: GT 1941 Batch, Box: Technical Training, Graduate Trainees 1941, 1944, 1945, TSA.

Notes: * The full names of the colleges listed are as follows:

    BHU: Engineering College, Banaras Hindu University, Banaras.

    Jadavpur: College of Engineering and Technology, Bengal (known as Bengal Technical Institute until 1929), located in Jadavpur near Calcutta.

    VJTI: Victoria Jubilee Technical Institute, Bombay.

only a select few colleges operating established courses in branches relevant to industry, such as electrical and mechanical engineering. All students listed in the table except N. K. Ganguly had obtained their degrees or diplomas in colleges that had been set up and were funded, wholly or partially, through private initiative. Indeed, the college at Jadavpur had been established in 1906 as a result of the Swadeshi movement's boycott of colleges run by the colonial government.[122]

It appears that B Class Apprentices were expected to become middle-level supervisory employees with the possibility of promotion,

---

[122] On Jadavpur, see Bagal, *Pramatha Nath Bose*, chapter X. On VJTI, see IIC *Report*, pp. 105–6. The moving force behind the setting up of BHU was the Congress leader Madan Mohan Malaviya. The engineering college of BHU was established in 1919. See Institute of Technology, Banaras Hindu University, 'Heritage', available at http://www.itbhu.ac.in/itbhu/heritage.shtml, accessed 9 June 2012.

while A Class students were groomed to attain higher managerial positions (as indicated by the cases of B. K. Mukherjee and V. M. Subramanian in the table). Much, however, depended on the individual and his performance during the course of training (and, naturally, after employment). For instance, although A Class apprentices were usually offered a higher starting salary if employed at the end of the course, a B Class apprentice could in theory be offered the same salary provided his performance in the course reached the standard of an A Class student.[123]

***

By all accounts the JTI helped the management successfully Indianize the works. As of 1938–9, 219 students had been trained in the JTI, of whom 202 were working in the company.[124] A recent study by Hiruyoki Oba and Hrushikesh Panda gives the number of covenanted personnel replaced by Indians in the period 1926–33 as eighty-five,[125] a significant number given that the total strength of covenanted personnel in 1925–6 had been 199.[126] It follows that considerable savings were effected, as Indians were less expensive than their foreign counterparts even at the managerial levels. Oba and Panda estimate that expenditure on wages decreased 'by about 50%', which 'thereby led to sizeable reductions in the average cost of production of steel'.[127]

The JTI played an important role in the continued healthy performance of TISCO in the interwar period. With the help of economic protection, TISCO successfully withstood foreign competition, while domestic demand grew as a number of ancillary steel-consuming industries came up in Jamshedpur (for example, Agricultural Implements Company; Tinplate Company of India Limited; Indian Steel &

[123] See correspondence in U. A. Prabhu file, Folder: 1938 Batch, Box: Technical Training, A&B Class Apprentices, 1937, 1938, 1939 Batches, TSA.

[124] This would include the students who belonged to the three-year programme before 1935, and the A and B Class apprentices thereafter. Directors' Report, in TISCO Annual Report for 1938–9, p. 8.

[125] Oba and Panda, *Industrial Development and Technology Absorption*, p. 73.

[126] Calculated from Statements no. 57 to 70, ITB 1926, pp. 115–19.

[127] Oba and Panda, *Industrial Development and Technology Absorption*, pp. 81–2.

Wire Products Ltd; Jamshedpur Engineering & Manufacturing Co).[128]
TISCO's share of the domestic steel market by volume was 37.3 per
cent in 1926; rose to 59.45 per cent (431,000 out of 725,000 tons) in
1932–3, despite a fall in overall demand in the Depression years; and
increased further to 66 per cent around the start of World War II.[129]
During the war, the company began to produce a wide variety of steel
products such as 'armour plates ... alloy, tool and special steels' and
'acid steels for turning out wheels, tyres and axles' for the railways.[130]
The JTI did not become the only entry route to TISCO for technical
experts—P. N. Mathur being an example of an experienced expert
being recruited directly—but it became essential to the company's
recruitment strategy. The availability of qualified Indian personnel at
a rate that the company could determine, through the JTI, was crucial
in cutting costs, convincing the government to continue the grant of
protection, and running the expanded works after the Greater Extensions had been carried out. In achieving this, the JTI also established
itself as one of the earliest industrial training facilities in the country. The three-year programme that it ran in its first phase, starting
in 1921, anticipated by at least a decade the introduction of degree
courses in mechanical and electrical engineering in most of the government engineering colleges.[131] This suggests that the two-year
graduate trainee programme of the JTI's second phase was an exceptionally advanced programme in the context of 1930s India.

***

The Tata Iron and Steel Company, perhaps the most important Indian
industrial enterprise in the interwar years, was not only built by
American engineers but also staffed by a group of technical experts
from various countries, of which the USA was the most prominent.
The culture of steel-making in the company (whose works were inaugurated in 1912) was defined by this multinational group of technical

---

[128] Ray, *Industrialization in India*, pp. 77–8, 89–90; ITB 1924, p. 74.

[129] Ray, *Industrialization in India*, pp. 87–8.

[130] Ray, *Industrialization in India*, p. 92.

[131] See Kumar, 'Colonial Requirements and Engineering Education', endnote 36
on p. 231 for the dates when these degrees were introduced at Poona, Sibpur, and
Madras.

managers and supervisors, men of physical strength, many of whom had learnt their skills through practical training. When they were joined by a number of Indians—who had attended universities in the USA, Germany, and Britain and been industrial apprentices in those countries—the culture of steel-making continued to emphasize the importance of practical experience, physical toughness, and quick-wittedness in the works.

The JTI was central to the process of Indianization that was undertaken in the interwar years, as the company sought to cut costs and obtain economic protection from the government. The graduates of the Institute were trained to replace foreign employees at the middle and higher levels of the works' personnel. The most important feature of the Institute's curriculum was its mix of theoretical learning and practical experience—students spent a substantial proportion of their training as apprentices in the TISCO works. Both at its inception, when it trained school-leavers, and from 1935, when it became a training programme for graduates in metallurgy and engineering, the JTI's management continued to prize physical fitness and practical experience in addition to academic achievement. These formed important criteria in the selection of students for the JTI and ultimately for jobs in the works. The resulting culture of industrial engineering, which privileged technical specialization and shop-floor experience, was very different from the gentleman-generalist paradigm of PWD engineers and the government-bureaucratic culture in which railway engineers (including those on company-run railways) operated.

This study of industrial experts in TISCO has contributed to the wider understanding of industrialization and Indianization in three significant ways. First, it has added a vital dimension to economic histories of Indian industry, in which technical experts have not received adequate attention. I have shown that in addition to economic protection, the ensuring of a steady supply of Indian experts was vital in enabling TISCO to effect savings and perform successfully through the interwar period.

Second, the chapter has highlighted the importance of extra-imperial networks and influences in the history of industrial engineers in India. The multinational team of experts who started up the works, American managers, machinery, and consulting engineers, and

Indians trained in America and Germany were all essential factors in the company's interwar performance.

Third, the chapter has shown that the Indianization of technical practitioners in interwar India occurred in diverse ways. In contrast to the case of government-employed engineers in the PWD and railways (as analysed in Chapters 3 and 4), the motivation for Indianization in TISCO was largely internal and not so much the result of external demands by the public or by politicians. The engineers of the PWD and state railways belonged to public services, the Indianization of which was actively monitored by Indian members of the central legislature. Their demands carried particular weight in the light of the Montagu–Chelmsford Reforms, which had specified Indianization as an important goal of government policy; yet in practice, government officials often resisted Indianization, and always proceeded cautiously in implementing it. The case of TISCO was different. While the Tariff Board—which decided upon the award of protection to specific industries—required the company to show some progress in Indianization, this was an incidental factor, the main point of the investigation being the company's ability to cut costs and its potential for success under protection. Indianization at TISCO was essentially driven from within. The company was owned by Indians; foreign experts had been employed mainly to help start up the works; Indian experts were considerably less expensive; and no quotas were set for Indians. As far as can be discerned from the correspondence and statements quoted here of the American General Managers such as Wells, Keenan, and Tutwiler, there was no systematic prejudice against employing Indian engineers for the TISCO works. Finally, the company did not rely on the colonial education system as a source of Indian experts, using instead its own Institute, the JTI, in which many of the graduate trainees were products of a small number of privately run technical/engineering colleges.

The historian Ramachandra Guha was recently quoted as saying that two of the things that keep India together are cricket and Bollywood.[1] We might add a third: engineer-worship. According to the All India Council for Technical Education (AICTE), there were 3,346 institutions in the country offering undergraduate degree programmes in engineering and technology as of March 2012. Their combined 'intake capacity' was listed at more than 1.47 million.[2] The number of graduates in engineering (including computer science) would have been of a comparable order of magnitude: historian Ross Bassett quotes an estimate that 220,000 such graduates entered the market in the year 2005–6. 'India,' he writes, 'is awash with engineers.'[3]

The sheer magnitude of the numbers might still be surprising to some, but the general trend they suggest shocks nobody. Where once the maharaja and the snake-charmer were stereotypes of the country, it is now engineers that the world thinks of when it thinks of India. The Indian character Asok in the popular *Dilbert* comic strip, an IIT graduate, illustrates this perfectly. A generation of children is growing up with one particular career in mind. William Dalrymple's *Nine*

[1] 'Two Binding Factors', *The Hindu* (online), 11 July 2015, available at http://www.thehindu.com/news/national/karnataka/two-binding-factors/article7409791.ece, accessed 18 July 2016.

[2] *Annual Report & Audited Accounts: 2011–2012* (New Delhi: All India Council for Technical Education, n.d.), p. 65.

[3] Ross Bassett, *The Technological Indian* (Cambridge, MA, and London: Harvard University Press, 2016), p. 304. Kindle edition.

Lives contains the story of a sculptor in Swamimalai in southern India whose ancestors have been casting bronze statues for seven hundred years. His son wants to be a computer engineer.[4]

Crucially, these understandings of Indian engineers revolve around a particular type and definition of engineering, namely proficiency in high-tech industries such as semiconductors, information technology, and computers in general.[5] But the cult of the engineer in India is much older than the recent, post-liberalization phenomenon involving information technology, software, and outsourcing. What were engineering careers like in the nineteenth and early twentieth centuries? How did engineers form a profession and raise their prestige in society?

This book explores the backstory of the phenomenon of the Indian 'techie', describing a burgeoning engineering profession in the first half of the twentieth century—a profession that was increasingly seeing itself as Indian and not a mere offshoot of the metropolitan (British) profession.

***

This has been the first extensive historical study of the engineering profession in India—a profession whose members numbered in the thousands, were central figures in government bureaucracies as well as industrial enterprises, and operated across the length and breadth of the subcontinent. Focusing on the period 1900–47, the book has explored the functions and the changing organization of engineers in three different areas: public works, railways, and private industry. Military engineers, who were a sizeable presence in public works and railways, were also included in this analysis. This was juxtaposed with a view of the rivalries within and the development of the engineering profession as a whole, reconstructed through an analysis of the role

---

[4] William Dalrymple, *Nine Lives: In Search of the Sacred in Modern India* (London, Berlin, and New York: Bloomsbury, 2009), chapter 7, 'The Maker of Idols'.

[5] For a discussion of the IT industry and its Nehruvian origins, see Dinesh C. Sharma, *Nehru: The Unlikely Hero of India's Information Technology Revolution*, NMML Occasional Paper, Perspectives in Indian Development, New Series 8 (New Delhi: Nehru Memorial Museum and Library, 2013).

of professional institutions. The contested process of change—especially Indianization—was in each case closely related to the multi-layered and continuously evolving colonial state in the interwar era of constitutional reforms. In the following section I present the main conclusions of the substantive chapters before identifying the broad conclusions that emerge from the book as a whole.

***

This book began by tracing the broad developments in the engineering profession through an analysis of the comparative importance of professional institutions in Britain and India. Two major conclusions were drawn. First, the profession grew in size over the period 1900–47, and the share of mechanical and electrical engineers—most of them working in private industry—grew considerably in comparison to that of civil engineers, who had dominated the profession in India before World War I. This is indicated in the membership of the London-based Institutions of Civil, Mechanical and Electrical Engineers. This shift in the composition of the profession reflected the government's new economic policies—more favourable than before to the growth of large-scale industries—which, in turn, were a result of the changing economic and political relationship between Britain and its Indian empire in the interwar period.

Second, the membership trends of the London institutions revealed another shift, one that was crucially important in all branches of engineering: Indianization. In all three London institutions, the percentage share of 'native' Indians among India-resident members increased considerably. The transformation, however, went beyond a mere increase in the number of Indian engineers. The creation of the Institution of Engineers (India) (IEI) in Calcutta in 1920, its recognition by the central and provincial governments, and its growth to prominence in the interwar period pointed to the formation of a nascent Indian identity in the engineering profession. Before World War I, engineers' identities had either been constructed as a part of the metropolitan/empire-wide engineering profession (through membership of the London institutions), or as specific to the elite government services (as in the case of the PWD Congresses founded in the twentieth century). By contrast, in the interwar period, key figures at the all-India and provincial levels of the IEI, whether Indian or British, consistently referred to the profession as being bound up with the cause of Indian economic development and industrialization.

Having established the overall trends of industrialization and Indianization in the trajectory of the engineering profession, the rest of the book identified the complexities and conflicts involved in these processes, and the ultimate limits of change in the profession. These features came to light through detailed studies of engineers in different fields of government and industrial work.

Beginning with the Public Works Department, I demonstrated that although measures were taken in response to demands for more equal opportunities for Indians, inequalities persisted, and the rate of Indianization was gradual. At the start of our period there were two separate engineering services, an Imperial Service (staffed by recruits from Britain) and a Provincial Service (recruiting in India), which differed in prestige and emoluments. In 1920, ostensibly to remove these distinctions, a new Indian Service of Engineers (ISE) was created, recruiting in India as well as Britain. Yet a new set of distinctions in allowances and privileges was introduced between India- and Britain-recruited engineers. Furthermore, new provincial services were created, below the ISE officers in the PWD hierarchy of each province, and it was in these services that most fresh opportunities were available for Indian engineers. In the elite ISE, Indian engineers approached a 50 per cent share in the late 1930s.

The misgivings of colonial officials and British policymakers about Indianization came to the fore when first the Roads and Buildings branch and later the Irrigation branch of the PWD were provincialized (in 1919 and 1935 respectively). This meant that recruitment of engineers would be the concern of the provincial governments rather than the Secretary of State in London—and, consequently, that most new recruits would be Indians. I showed that opponents of these changes suggested that Indians were deficient in courage, a sense of responsibility, and general character rather than in technical aptitude. During a 1935 debate in the House of Commons on recruitment to the Irrigation service, doubts were cast also on the integrity of Indian engineers, their ability to combat corruption, and their impartiality in the distribution of irrigation water.

These criticisms were directly related to the culture of public works engineering. The ideal PWD engineer was supposed to be a generalist and a gentleman—more specifically a British gentleman—whose character was as important as his technical ability. The training and recruitment of engineers, which had long been under the close control of the India Office in London, was designed to this end—as illustrated by the curriculum and lifestyle of students at the PWD-focused Cooper's Hill College near London. The near-monopoly of Cooper's Hill graduates on PWD (Imperial Service) appointments until the College's closure in 1906, the ubiquity of those graduates in the upper ranks of the ISE through to the 1930s, and the India Office's efforts to attract applicants with a similar pedigree after 1906 ensured that this culture pervaded the PWD. My more general point is that a good way to understand the public works engineer in our period is as a particular type of colonial administrator.

If Indianization in the PWD was related to provincialization, in the railways it was closely associated with demands for the *nationalization* of company-run lines. Historians such as Ian Kerr have stressed that the demands of nationalist politics drove the history of the railways in the years 1900–47. In this book I have concentrated on analysing the response of the colonial government, in particular their resistance to change in the Superior Services (officer positions) of the railways. Although the Acworth Committee (1920–1) decided in favour of nationalizing the company-run railways, it did so only by a slender majority (the Chairman's casting vote). The Committee's minority and other critics opposed nationalization on the grounds that efficiency would suffer, that democracy did not lend itself to good railway management, and that the organization would become inflexible. Further, nationalization did not necessarily lead to greater Indianization as was expected.

Following the recommendations of the Islington Commission (1912–15) and Lee Commission (1923–4), the government had set a target of recruiting fresh officers in the ratio of three Indians to every European, with the aim of achieving a 50–50 composition overall as quickly as possible. These recruitment targets, which were to apply to company-run as well as state-run railways, were consistently missed, and it was only in the late 1930s that Indians made up one half or more of the Superior Services. Even this result was due not only to the recruitment of more Indians, but also to the departure or retrenchment of Europeans during the Depression years and after the start of World War II. In demonstrating this, I support Daniel Headrick's claim that Indianization was partly the result of external factors. While Headrick's argument was made for railway employees as a whole, my statistical analysis has confirmed that it also applies for the specific case of Superior Service officers.

My analysis has also shown that Indianization was pursued selectively, in such a way as to ensure continuity in the top echelons of the railway bureaucracy, while allowing a notable degree of change in the middle and lower levels. New facilities set up in the interwar period for the training of Indians concentrated upon the Traffic and Mechanical Engineering departments of the railways, while no fresh measures were introduced to train Indians for the more prestigious and powerful civil engineering positions. Correspondingly, the proportion of Indians among fresh recruits to the engineering department was lower than that in the other departments of the railways. As in the case of the PWD, Indianization had its vocal opponents and sceptics. The arguments here, however, were slightly different: in the place of integrity and impartiality, 'efficiency' was invoked. The railways were to be run on business principles, and British engineers with practical experience on railways in their home country were considered better candidates to run the Indian railways efficiently. Additionally, the spectre of sabotage was raised: it was argued that the railways, being the strategic lifeline of the Empire in India, required the dominant presence of 'loyal' (that is, European) engineers.

While the key to the recruitment of Indians in the PWD and railways lay with the colonial government and its engineering colleges in India, private industry had other options, as shown by the case of the Tata Iron and Steel Company (TISCO), the premier heavy industrial enterprise in interwar India. The initial dependence of TISCO on government patronage and the grant of economic protection notwithstanding, the company's recruitment and training of its technical personnel had an important extra-imperial component. In the beginning the company imported experts from several countries, primarily the USA; a generation of Indians trained in Germany and the USA then occupied important positions in the company; and after World War I, the company successfully set up and ran its own postgraduate training school, the Jamshedpur Technical Institute (JTI), to train Indians for supervisory positions in the works. Many of the entrants to the JTI (and thence to TISCO) were graduates of colleges associated with nationalist leaders (for example, the Banaras Hindu University, and the Bengal Engineering College, Jadavpur).

The importance of relationships beyond the confines of Empire has recently been highlighted by historians, as in Ross Bassett's study of a small group of Indians who went to the Massachusetts Institute of Technology to train as engineers and technologists, funded largely by princely rulers from the Kathiawar region, returning to set up Swadeshi industries in India or participate in nationalist agitations.[6] My study of TISCO has revealed the features of another important tribe of America-trained Indian engineers—experts who did not necessarily participate directly in the nationalist movement or aim to set up completely indigenous industries, but worked in a large-scale industry which, while owned by Indian capital, maintained a close working relationship with the colonial government.

The experience of TISCO differed in other ways from that of the government services. Indianization here was largely an internal process, driven more by concerns of economy than by political demands. Further, a different work culture prevailed in place of the ideal of the gentleman engineer. The steel works were a place for unpretentious specialists with practical knowledge, stamina, and the willingness to work with their hands. Elements of this culture, which first took shape under the early generations of foreign experts who had learnt their skills by apprenticeship in the steel mills of their own countries, were continued and institutionalized in the subsequent generation of schooled engineers (as seen in the training programme at the JTI).

***

Taken together, these specific findings reveal several broad patterns that demonstrate the utility of the approach adopted in this book.

To begin with, this book has stressed the need for placing practitioners at the centre of the history of science and technology in India.

---

[6] Ross Bassett, 'MIT-Trained Swadeshis: MIT and Indian Nationalism, 1880–1947', *Osiris*, vol. 24, no. 1 (2009): 212–30.

Focusing thus on engineers and their professional concerns has led to a deep engagement with the bureaucratic structure of government services and large-scale private industries, and their systems for the recruitment and training of technical experts. This has enabled us to see science and technology in the period 1900–47 not just as a harbinger of modernity or as a knowledge system, but also as the exercise of functions of fundamental administrative and economic importance. The practice of technology was as much a question of career, status, and Indians' desire for a role in government as it was about 'knowledge' or the building of a nation.

This study has deliberately focused on the twentieth century, to which a relatively small section of the historiography of science and technology in India is devoted. I have sought to take into account, and contribute to our understanding of, the fluid nature of the Indian economy, polity, and society in this period, and the multiplicity of competing but interdependent actors and interests. By examining the role of engineers in these transitional decades between Victorian colonialism and Indian Independence, this book has questioned the utility of essentialized categories like 'colonialism' and 'nationalism' in understanding the history of science in India. As this book has shown, it is neither useful nor accurate to view the science and technology of the state as necessarily 'colonial', to be placed in opposition to a nationalist movement that is considered as operating entirely outside the state machinery. On the contrary, the Indian politicians in the provincial and central legislatures were themselves part of the state; and their demands for Indianizing the public works and railway engineering services amounted to their wanting more Indians to work for the state. My study of TISCO's technical experts has shown a similar blurring of categories. The company cannot easily be bracketed as loyalist or nationalist.[7] It maintained a close working relationship with the colonial government, supplying it with steel rails during the Great War and obtaining economic protection from it in the 1920s and 1930s. Yet TISCO circumvented (or had little use for) the main

[7] On a related note, see Vinay Bahl's labour history of TISCO, one of the themes which is the company's successful maintaining of relations both with the colonial government and with nationalist leaders, as it tackled a series of major strikes. Vinay Bahl, *The Making of the Indian Working Class: A Case of the Tata Iron and Steel Company, 1880–1946* (New Delhi, Thousand Oaks and London: Sage Publications, 1995).

state-run engineering colleges in sourcing technical personnel for its works in Jamshedpur. Indeed, the engineering profession at large, as represented by the Institution of Engineers (India), evolved an Indian identity while remaining wedded to the colonial state. The creation of the Institution followed the recommendation of a government-appointed body, and had the ultimate official recognition in the form of its 1935 Royal Charter. Yet the Institution was encouraging of Indi-anization and the promotion of industries, and the engineers who occupied leadership roles within the organization increasingly articulated their belief in a form of economic nationalism.

I have also emphasized in this book the need to pay close attention to the two-way relationship between science/technology and the heterogeneous and evolving colonial state in our period. On the one hand, I have shown how the careers and professional opportunities of engineers were impacted by the constitutional reforms occurring in 1919 and 1935. Indian members of the expanded Legislative Assembly were instrumental in pushing the Indianization agenda in the railways: the setting of quotas for Indian recruitment under the Lee Commission and the subsequent annual reporting of progress in that direction could probably not have occurred in the India of earlier decades. The newly empowered provincial governments (which were in charge of Industries from 1919) also played a role in the case of private industrial enterprises like TISCO. The Jamshed-pur Technical Institute was (in its early years) funded partially by the governments of Bengal and Bihar and Orissa, and continued to have a relationship with provincial technical colleges into the 1940s, accepting students from them for its trainee schemes. On the other hand, I have also shown that the actual implementation of constitutional reforms was a contingent process, whose particular form often depended on the state's understanding of the role of its engineer (and other) officers. When Roads and Buildings was provincialized after World War I, it raised the question: to whom would public works engineers be responsible—to the Secretary of State, to the Government of India, or to the government of the province in which they were working? In the event, while fresh recruits were made responsible to provincial ministers, the retention of the existing ISE officers under their old terms was considered essential to maintain continuity. Similarly, when the Irrigation branch was provincialized after 1935, the Secretary of State reserved the right to make appointments to it in exceptional cases. Provincialization, in both cases, came with caveats and qualifications.

The present study has confirmed that Indianization was a ubiquitous theme in the years 1900–47, relevant in all the sectors in which engineers worked. Debates, whether on the provincialization of the PWD's Irrigation branch or the nationalization of company-run railways, hinged on the participants' views on Indian engineers and whether they could successfully replace Europeans. Examining the process of Indianization has further shown that the question of race was central to the experience of professional engineers in our period, even if it was not always referred to directly. The initial Imperial/ Provincial distinction, the multiple Royal Commissions on Indianization, the prescribing of quotas for recruitment in Britain and India (in some cases) and of Europeans and Indians (in others) were all manifestations of this fact. The debates on Indianization give us a clue as to how race was understood in relation to technical practitioners. The race of an engineer was commonly associated with many qualities: gentlemanliness, integrity, and courage (in the PWD), 'efficiency' and loyalty to the colonial regime (in the railways), but seldom directly with technical competence.

Further, the book has demonstrated the utility of a little-used approach to the study of industrialization in interwar India. As my study of the steel experts of TISCO suggests, focusing on industrial engineers can help us understand various facets of the interwar growth of large-scale industry, including the state of industrial education in the country, the role of international linkages in the flow of experts and machinery, the relationships of industries with the colonial government on the one hand, and with provincial governments on the other (Industries being a transferred/provincialized subject in the interwar period). The experts who carried out the daily operations of TISCO were at the heart of its activities: economic protection and the state of international trade in the interwar period doubtless played an important part in TISCO's success, but would have been of little use if the company had not been able to find a way to procure technical experts. In turn, such studies, seen in conjunction with studies of government services, tell us a great deal about the differing experiences of engineers across sectors, and about the increasingly diverse range of activities that constituted engineering in India over the course of the interwar period.

In summary, structuring this study of engineers around the themes of Indianization and industrialization has not only revealed important aspects of those two processes, but also served to situate the history of engineers in India within the broader framework of Indian history.

FURTHER RESEARCH

The story of engineers, of course, does not end in 1947. As we saw at the beginning of this chapter, a degree in engineering is one of the most sought-after qualifications in twenty-first-century India. In particular, there is an emphasis on computer science, and electronics and communications engineering. There has been a corresponding rise in scholarly interest in the current state of the engineering profession. An emerging literature dwells on India's experience in the world of information technology, software, and outsourcing.[8] Practitioners and scholars have also made perceptive contributions to the debate on how the engineering profession has developed in recent decades. Two important arguments that have been made are that the prestige of (and hence competence in) civil engineering has fallen relative to the newer branches of engineering,[9] and that engineering education in elite institutions and research and development in engineering have not been grounded in the needs of Indian society.[10]

There has been a particular flowering of interest (from disciplines other than history) in the experience of students at the renowned Indian Institutes of Technology (IITs) and the role of these institutions in Indian society. A recent study of IIT Madras by anthropologist Ajantha Subramanian discusses the relationship between caste and other forms of social capital and the idea of the IITs as a 'meritocracy'. Specifically, Subramanian 'argue[s] that the IIT graduate's

[8] Carol Upadhya, *Reengineering India: Work, Capital, and Class in an Offshore Economy* (New Delhi: Oxford University Press, 2016); Dinesh C. Sharma, *The Outsourcer: The Story of India's IT Revolution* (Cambridge: MIT Press, 2015).

[9] Shirish B. Patel, 'Why Flyovers Will Fall: Decline of the Civil Engineering Profession in India', *Economic and Political Weekly*, vol. 51, no. 20 (14 May 2016): 32–6.

[10] Milind Sohoni, 'Engineering Teaching and Research in IITs and Its Impact on India', *Current Science*, vol. 102, no. 11 (10 June 2012): 1510–15.

status depends on the transformation of privilege into merit, or the conversion of caste capital into modern capital'.[11] Focusing on IIT Kanpur and against the backdrop of reservations for students belonging to historically disadvantaged groups, Odile Henry and Mathieu Ferry seek, in an ongoing study, to investigate the popular belief that the IITs are 'a vehicle for social mobility based only on the students' skills'.[12] Sociologist Roland Lardinois has begun an investigation into the 'coaching' industry that has sprung up in the Rajasthani town of Kota, essentially to prepare students for the IITs' all-India Joint Entrance Exam (JEE).[13]

Yet, while there is great interest in the state of engineering today, there is plenty more to be understood about the role of engineers between the twenty-first century and the interwar period that this book focuses on. To begin with, the relationship between engineers and the state in the early post-Independence decades was of great importance. The first independent governments under Nehru, in their policies on technical education (broadly defined), inherited elements of the approaches both of their nationalist forebears and of the colonial government. In the process, they began the trend of privileging elite forms of technical education (especially engineering degrees) over non-elite forms such as vocational or in situ industrial education.[14] In the particular context of the education of professional engineers, Kim Sebaly has studied the origins of the IITs, which became the most prestigious engineering institutions in the post-Independence

[11] Ajantha Subramanian, 'Making Merit: The Indian Institutes of Technology and the Social Life of Caste', *Comparative Studies in Society and History*, vol. 57, no. 2 (2015): 291–322. Quoted text from 'Abstract' on p. 322.

[12] Odile Henry and Mathieu Ferry, 'Sociology of UG students at IIT Kanpur: Education-Path and Placement according to Admission Status', presentation at the mid-term workshop of the ENGIND ('Engineers and Society in Colonial and Post-Colonial India') project under the French National Research Agency, New Delhi, 11–12 January 2016.

[13] Roland Lardinois, 'Coaching the Masses, Creaming the Elites: The Education Market at Kota (Rajasthan)', presentation at mid-term workshop, ENGIND, New Delhi, 11–12 January 2016.

[14] See Aparajith Ramnath, 'Breaking Free: Technical Education Policy in India Immediately before and after Independence' (MSc dissertation, University of Oxford, 2007).

years.[15] Further research would study how these education facilities were related to the employment opportunities in the 1950s and 1960s, not only in government services and private industries, but also in the new public sector heavy industries that were inaugurated in this period.

Among the engineers whose careers peaked after Independence were A. N. Khosla and Kanwar Sain, both of whom feature in Daniel Klingensmith's study of large dam projects and the discourse of development. Klingensmith indicates the importance of reformist social and educational movements like the Arya Samaj, which helped create a 'modernist, professionalized Punjabi middle class', and gave rise to individuals like Khosla and Sain.[16] The phenomenon of a new professions-oriented class in Punjab is vividly described in *Punjabi Century*, a well-known memoir by Prakash Tandon (himself the son of a PWD engineer),[17] and also referred to in a recent memoir by Jagman Singh (an engineer on the Bhakra project in Punjab).[18] Research exploring whether similar processes occurred in other regions of India would be useful in further understanding how ideas of nation-building impacted the careers of engineers in post-Independence India.

An exciting direction in which the themes presented in this book may be extended is the role of engineers in twentieth-century Indian industrialization. This book has shown in detail the history of engineers and international networks in interwar TISCO. TISCO was, of course, an exceptionally successful industrial enterprise, but there is reason to believe that the experience of its engineers was in many ways representative of trends in large-scale industry. There is evidence, for instance, to suggest that similar features (foreign experts in the beginning, Indians trained in the USA) were to be found in other emerging industrial enterprises. An excellent instance is the

[15] Kim Patrick Sebaly, 'The Assistance of Four Nations in the Establishment of the Indian Institutes of Technology, 1945–1970' (PhD thesis, University of Michigan, 1972).

[16] Daniel Klingensmith, *'One Valley and a Thousand': Dams, Nationalism, and Development* (New Delhi: Oxford University Press, 2007), chapter 5. The quoted text is from p. 226.

[17] Prakash Tandon, *Punjabi Century: 1857–1947* (London: Chatto and Windus; Toronto: Clarke, Irwin and Co., 1961).

[18] Jagman Singh, *My Tryst with the Projects Bhakra and Beas* (New Delhi: Uppal Publishing House, 1998), esp. p. 59.

Kirloskar group, which manufactured agricultural implements at its industrial township near Poona. As in the case of TISCO, a strong American influence operated here. The founder, Laxmanrao Kirloskar (a former instructor at Bombay's Victoria Jubilee Technical Institute) had been a subscriber to *American Machinist* in his youth, and sent his son Shantanu (who later headed the company) to the USA to study engineering at the MIT.[19] Stefan Tetzlaff is engaged in work that is likely to tell us more about the role of international networks in industrialization, through case studies of the Tata Engineering and Locomotive Company (TELCO), which built trucks in collaboration with Mercedes Benz of Germany, and Premier Automobiles Ltd, which collaborated with Chrysler of the USA and later with Fiat of Italy.[20] My own developing work on the history of the Hindustan Aircraft Limited (est. 1940) seeks to understand the interplay between local, colonial, and extra-colonial factors in the development of the aircraft industry in India, during and after World War II.[21] All these approaches lend themselves, I believe, to a new way of understanding twentieth-century industrialization in India. Recent books on the subject focus on how entrepreneurship was related to industrialization,[22] and how Indian capitalists interacted with the postcolonial state;[23] both use the notion of 'late' (or 'late, late') to describe Indian industrialization. Just as this book has suggested for the interwar period, zeroing in on the training and flow of engineers can lead to novel ways

[19] S. L. Kirloskar, *Cactus & Roses: An Autobiography* (Pune: C.G. Phadke, 1982). For a recent treatment of the Kirloskar case, see Ross Bassett, *The Technological Indian*, which also describes other important industrial enterprises begun by Indians, including the Paisa Fund Glass Works, associated with the engineer Ishwar Das Varshnei, and a chemical factory started by Devchand Parekh and his brother in Bhavnagar.

[20] Stefan Tetzlaff, 'State Policy, Technical Cooperation and Manpower Requirements: The Case of Indian Automotive Engineering around the Mid-20th Century (c. 1947–1970)', presented at midterm workshop, ENGIND, New Delhi, 11–12 January 2016.

[21] Aparajith Ramnath, 'International Networks and Aircraft Manufacture in Late-Colonial India: Hindustan Aircraft Limited, 1940–47', IIM Kozhikode working paper no. IIMK/WPS/205/HLA/2016/17, July 2016.

[22] Sumit K. Majumdar, *India's Late, Late Industrial Revolution: Democratizing Entrepreneurship* (Cambridge, UK, and New York: Cambridge University Press, 2012).

[23] Vivek Chibber, *Locked in Place: State-Building and Late Industrialization in India* (New Delhi: Tulika Books, 2004).

of understanding the postcolonial period, ways that need not assume a standard model of industrialization.[24]

Finally, some of the emerging work I have cited above stems from a new international research project (of which I am a member) that is due to report its findings in 2017. Titled 'ENGIND: Engineers and Society in Colonial and Post-Colonial India', this collective under the aegis of the French National Research Agency[25] has scholars working on an eclectic mix of contemporary and historical topics that promise to further our understanding of engineering in India.[26]

***

A further understanding of the vocation of engineering after Independence, and especially in the last few decades, must thus await the fruits of the ongoing efforts of various scholars, and there is much to look forward to. But the growing importance of engineering after Independence would not have been possible without the base that existed in 1947, and any understanding of the profession must rest on how it came into being. That base was a largely united albeit diverse profession, the bedrock of which was a mix of public works, railway, and industrial engineers, both Indian and European. Crucially, as this book has argued, it was a profession that had begun to become Indian.

[24] This does not invalidate the enquiries of scholars who wish to understand the gap between the Indian state's goals and performance in relation to industrialization; what it does is offer an alternative, and potentially illuminating, perspective.

[25] The website of the project, coordinated by Dr Vanessa Caru, is http://engind.hypotheses.org/ (accessed 15 October 2016).

[26] In addition to the studies by Henry/Ferry, Lardinois, and Tetzlaff cited earlier, projects with a historical aspect include N. C. Narayanan, Poonam Argade, and D. Parthasarathy, '"Dis(coursing) Water": Dams and Changing Engineering Paradigms in Western India'; Vanessa Caru, 'The Creation of a "corps d'état"? Indian and European engineers of the Bombay Public Works Department (1860's–1940's)'; and Aparajith Ramnath, 'Industrial Experts in the Age of Indianisation: The European Engineering Firms of Calcutta, 1914–47', all contributions to the mid-term workshop of the ENGIND project, New Delhi, 11–12 January 2016. I am grateful to my fellow researchers for permission to cite their work in progress.

PRIMARY SOURCES

Archival/Unpublished Sources

ASIA, PACIFIC AND AFRICA COLLECTIONS (APAC), BRITISH LIBRARY

Hall, G. F. 'All in the Day's Work'. Typescript, 2 volumes, 1947. APAC: Mss Eur D569/1 and D569/2.

Herbert Fagent Merrington Papers. APAC: Mss Eur D1229/1 and D1229/2.

Regulations Governing Appointments of Engineers, with Selection Committees' Reports. APAC: IOR/L/PWD/5/29.

INSTITUTION OF CIVIL ENGINEERS (ICE) ARCHIVES, LONDON

'The Institution of Civil Engineers: Advisory Committees'. Accession Number: 185/01.

Memorandum on 'Election of Colonial Representatives on the Council'. Accession Number: 185/01.

J. W. Mackison to A. A. Biggs, 30 March 1926, supporting proposal for a Local Association of the ICE in India. Accession Number: 185/03.

R. D. T. Alexander to Secretary of the ICE, 26 April 1926, part of the ICE's official correspondence of 1926 considering the creation of a local centre in India. Accession Number: 185/03.

INSTITUTION OF ENGINEERING AND TECHNOLOGY (IET) ARCHIVES, LONDON

Meares, John Willoughby. 'At the Heels of the Mighty: Being the Autobiography of "Your Obedient Humble Servant"'. Typescript, 1934. Accession Number: SC 169/1/1.

TATA STEEL ARCHIVES, JAMSHEDPUR (TSA)

'A Chart Showing the Organisation of the Staff, Both Administrative and Departmental with the Monthly Expenditure of Each Dept.', undated diagram [estimated interwar period].

Box 51: Private Papers, R. Mather.

Box 69: 'Private Papers: P. N. Mathur'.

Box: General Manager's Correspondence, 1909.

Box: Technical Training, A&B Class Apprentices, 1937, 1938, 1939 Batches.

Box: Technical Training, Graduate Trainees 1941, 1944, 1945.

Box: Technical Training, T1/A1 (1921–60), A1 (1962–4, 1965–8, 1966–84).

Box: Technical Training: Bihar and Orissa Apprentices, 1927–33 Batches.

Sahlin, Axel. 'Personal Impressions of India: Written for his friends by Axel Sahlin: January 15th to April 21st, 1908'. Booklet consulted at the TSA.

Books and Articles

'Address by the President Edward Hopkinson, Esq., DSc, M.P.', October 1919. In IMechE *Proceedings*, October–December 1919, pp. 631–58.

'Alexander Jardine, DSc'. Obituary appearing in *ICE Proceedings*, vol. 48, no. 4 (April 1971): 723–4.

Addis, A. W. C. *Practical Hints to Young Engineers Employed on Indian Railways*. London and New York: n.p., 1910.

Ashby, Lillian (with Roger Whately). *My India: Recollections of Fifty Years*. Boston: Little, Brown and Company, 1937.

Bayley, Victor. *Nine Fifteen from Victoria*. London: Robert Hale & Company, 1937.

Christie, J. M. 'The Manufacture of Oxygen and Its Use for Welding and Metal-Cutting'. IMechE *Proceedings* (December 1916): 889–94.

Curtis, L., ed. *Papers Relating to the Application of the Principle of Dyarchy to the Government of India: To Which Are Appended the Report of the Joint Select Committee and the Government of India Act, 1919, With an Introduction by L. Curtis*. Oxford: Clarendon Press, 1920.

Fermor, L. L. 'Thomas Henry Holland. 1868–1947'. *Obituary Notices of Fellows of the Royal Society*, vol. 6, no. 17 (November 1948): 83–114.

Hayavadana Rao, C., ed. *The Indian Biographical Dictionary 1915*. Madras: Pillar & Co, n.d.

The Hindusthan Association of America [New York City]. *Education in the United States of America: For the Guidance of the Prospective Students from India to the United States*, bulletin no. 1, 2nd and revised edn. New York City: n.p., 1920. British Library Shelfmark: General Reference Collection 8385.a.14.

Historical Retrospect of Conditions of Service in the Indian Public Works Department (All India Service of Engineers). Private pamphlet (c. 1925) in the Secretary of State's Library Pamphlets, vol. 72, T. 724. APAC: P/T724.

Institution of Engineers (India) Yearbook 1964–65. Calcutta: Institution of Engineers, 1965. British Library Shelfmark: General Reference Collection P. 621/409.

Institution of Engineers (India): Bye-Laws effective 5 May 2012. Available at http://www.ieindia.info/PDF_Images/Bylaws/ByeLaws.pdf. Downloaded 2 July 2012.

'Jo Hookm'. *The Koochpurwanaypore Swadeshi Railway*, 2nd edn. Calcutta and Simla: Thacker, Spink & Co, and Bombay: Thacker & Co Ltd, c. 1921.

Keenan, John L. (with Lenore Sorsby). *A Steel Man in India*. New York: Duell, Sloan and Pearce, 1943.

Khalidi, Omar, ed. *Memoirs of Cyril Jones: People, Society and Railways in Hyderabad*. New Delhi: Manohar Publications, 1991.

Kirloskar, S. L. *Cactus & Roses: An Autobiography*. Pune: C. G. Phadke, 1982.

Kunzru, Hirday Nath. *The Public Services in India (Political Pamphlets—II)*. Allahabad: Servants of India Society, 1917.

LeMaistre, C. 'Summary of the Work of the British Engineering Standards Association'. *Annals of the American Academy of Political and Social Science*, vol. 82, 'Industries in Readjustment' (March 1919): 247–52.

Leslie, Bradford. 'The Erection of the "Jubilee" Bridge Carrying the East Indian Railway across the River Hooghly at Hooghly'. *Minutes of the Proceedings of the Institution of Civil Engineers*, vol. 92 (1888): 73–96 (discussion: pp. 97–128; correspondence: pp. 129–41).

Maclean, C. D., ed. *Manual of the Administration of the Madras Presidency: In Illustration of the Records of Government & the Yearly Administration Reports*, reprint, 3 volumes. New Delhi: Asian Educational Services, 1987 [1885–93].

Mahalanobis, P. C. 'Sir Rajendra Nath Mookerjee: First President of the Indian Statistical Institute 1931–1936'. *Sankhyā: The Indian Journal of Statistics*, vol. 2, part 3 (1936): 237–40.

Molesworth, Guilford L. *Indian Railway Policy*. Indian Railway Series No. 1, compiled and edited by Faredun K. Dadachanji. Bombay: F. K. Dadachanji, 1920.

Napier, Philip. *Raj in Sunset*. Ilfracombe, Devon: A. H. Stockwell, 1960.

Obituary for Sir Robert Swan Highet, CBE, 1859–1934. *Minutes of the Proceedings of the Institution of Civil Engineers*, vol. 240, (1935): 787.

Obituary, 'Sir Thomas Guthrie Russell'. *ICE Proceedings*, vol. 31, no. 1 (May 1965): 126–7.

Reddy, D. V. *Inside Story of the Indian Railways: Startling Revelations of a Retired Executive*. Madras: M. Seshachalam, 1975.

Shah, K. T. *Public Services in India (Congress Golden Jubilee Brochure—7)*. Allahabad: All India Congress Committee, 1935.

Singh, Jagman. *My Tryst with the Projects Bhakra and Beas*. New Delhi: Uppal Publishing House, 1998.

Strong, Joseph F. 'On the Apparatus Used for Sinking Piers for Iron Railway Bridges in India'. *Proceedings of the Institution of Mechanical Engineers*, vol. 14, no. 1 (1863): 16–33 (including discussion).

Visvesvaraya, M. *Memoirs of My Working Life*. Bangalore: M. Visvesvaraya, 1915.

Wakefield, G. E. C. *Recollections: 50 Years in the Service of India*, pp. 27–8. Lahore: Civil and Military Gazette Ltd, 1942. Courtesy of the Centre of South Asian Studies archives, Cambridge.

Journals/Periodicals

Aberdeen Daily Journal, The

Engineer's Journal and Railway, Public Works, and Mining Gazette, of India and the Colonies

Indian Engineering

Indian Railway Gazette

Journal of the Institution of Engineers (India)

Lists of Members, Institutions of Civil/Mechanical/Electrical Engineers

Minutes of Proceedings of the Punjab PWD [Engineering] Congress

Minutes of the Proceedings of the Institution of Civil Engineers

Minutes of the Proceedings of the Institution of Mechanical Engineers

New York Times, The

Observer, The

Railway Gazette

TISCO Annual Reports (Directors' Reports)

TISCO News

TISCO Review

Transactions of the Indian Engineers' Association

Official Publications

SERIALS

(Hansard) House of Commons Parliamentary Papers, via ProQuest.

Hansard House of Commons Debates

Hansard House of Lords Debates

India Office List

PWD Classified List and Distribution Return of Establishment. APAC: IOR/V/13 Series.

Railway Board Reports
Statement Exhibiting the Moral and Material Progress and Condition of India
Thacker's Indian Directory

OTHER OFFICIAL PUBLICATIONS

'Regulations Governing Appointments of Engineers, with Selection Committees' Reports, 1905–1937'. APAC: IOR/L/PWD/5/29.

Administration and Progress Report of the Chief Engineer, United Provinces Public Works Department, Buildings and Roads Branch, For the year 1934–35. Allahabad: Superintendent, Printing and Stationery, United Provinces, 1936. Addressed by Chief Engineer Chhuttan Lal, to the Secretary to Government (United Provinces PWD, Roads & Buildings). APAC: IOR/V/24/3304.

Annual Report & Audited Accounts: 2011–2012. New Delhi: All India Council for Technical Education, n.d.

Appendix H: 'Scientific and Technical Societies', *East India (Industrial Commission). Report of the Indian Industrial Commission 1916–18.* Cmd. 51, London: HMSO, 1919, pp. 385–7. [IIC Appendix H.]

East India (Civil Services in India). Report of the Royal Commission on the Superior Civil Services in India. Cmd. 2128, London: HMSO, 1924. [Lee Commission Report.]

East India (Constitutional Reforms). Report on Indian Constitutional Reforms. Cd. 9109, London: HMSO, 1918. [Montagu–Chelmsford Report.]

East India (Railway Committee, 1920–21). Report of the Committee Appointed by the Secretary of State for India to Enquire into the Administration and Working of Indian Railways. Cmd. 1512, London: HMSO, 1921. [Acworth Committee Report.]

East India (Railways). Report on the Administration and Working of Indian Railways. By Thomas Robertson, C. V. O., Special Commissioner for Indian Railways. Cd. 1713, London: HMSO, 1903.

Extracts from the Debates in the Indian Legislature on Railway Matters: Delhi Session—January to March 1926. Calcutta: Government of India Central Publication Branch, 1926. APAC: IOR/V/25/720/44.

Extracts from the Debates in the Indian Legislature on Railway Matters: Delhi Session—January to March 1927. Calcutta: Government of India Central Publication Branch, 1927. APAC: IOR/V/25/720/45.

First Report of the Indian Tariff Board Regarding the Grant of Protection to the Steel Industry. Place and publisher illegible: [1924]. Available at Digital Library of India, www.dli.ernet.in., accessed 2 January 2012.

Government of India Act 1935. 26. Geo. 5. Ch. 2.

Government of India. *A Bill to Make Further Provision for the Government of India* (1934–5). House of Commons Parliamentary Papers. 25 Geo. 5.

'The Indian Industrial Commission: Its Report Summarised'. In *East India (Industrial Commission): Report of the Indian Industrial Commission, 1916–18.* Cmd. 51, London: HMSO, 1919, pp. 1–4.

Indian Industrial Commission 1916–18. *Report.* Calcutta: Superintendent Government Printing, India, 1918. [IIC *Report.*]

Indian Industrial Commission. *Minutes of Evidence 1916–17,* vol. 1: *Delhi, United Provinces and Bihar and Orissa.* Calcutta: Superintendent Government Printing, 1917.

Indian Railway Conference Association. *Proceedings of the Conference of Railway Delegates Assembled at Simla, October 1920.* Simla: Government Central Press, 1920. APAC: IOR/V/25/720/42.

Indian Railways Conference Association 1928. *Agenda and Proeedings of The Electrical Section: Meeting No. 1* (place and date of publication not given). British Library shelfmark: W 4026.

Indian Tariff Board. *Evidence Recorded During Enquiry into the Steel Industry,* vol. I: *The Tata Iron and Steel Company.* Calcutta: Superintendent Government Printing, India: 1924. [ITB 1924.]

Indian Tariff Board. *Evidence Recorded During Enquiry Regarding the Grant of Supplementary Protection to the Steel Industry.* Calcutta: Government of India Central Publication Branch, 1925. [ITB 1925.]

Railway Board. *History of Services of the Officers of the Engineer and State Railway Revenue Establishments, Corrected to 1st July 1910.* Calcutta: Superintendent Government Printing, India, 1910. APAC: IOR/V/12/67.

Report of the Indian Fiscal Commission, 1921–22. Simla: Superintendent Government Central Press, 1922.

Royal Commission on the Public Services in India. Report of the Commissioners, vol. 1. Cd. 8382, London: HMSO, 1917. [Islington Commission Report.]

The Applications Received by the Indian Tariff Board from the Tata Iron and Steel Company, Limited, and Other Companies in Connection with the Statutory Enquiry Regarding the Grant or Continuance of Protection to the Steel Industry in India after the 31st March 1927, Together with the Questionnaires Issued by the Board and Other Relevant Papers. Calcutta: Government of India Central Publication Branch, 1926. [ITB 1926.]

The Legislative Assembly Debates (Official Report), vol. 3, p. 46. Delhi: Manager of Publications, 1942. APAC: Microfilm IOR Neg 14795.

SECONDARY SOURCES

Books, Articles, Presentations, Working Papers

'Life and Work of Sir Rajendra Nath Mookerjee: The Inaugural-President of The Institution of Engineers (India)'. In *The Institution of Engineers (India), Diamond Jubilee 1980: Souvenir*, pp. 35–7. Calcutta, 1980.

'Stewart-Murray, Katharine Marjory, Duchess of Atholl'. *Concise Dictionary of National Biography*. Oxford University Press, March 1992. Available at http://www.knowuk.co.uk., accessed 29 January 2011.

'Turnour, Edward, Sixth Earl Winterton and Baron Turnour'. *Concise Dictionary of National Biography*. Oxford University Press, March 1992. Available at http://www.knowuk.co.uk, accessed 12 June 2012.

'Two Binding Factors'. *The Hindu* (online). 11 July 2015. Available at http://www.thehindu.com/news/national/karnataka/two-binding-factors/article7409791.ece, accessed 18 July 2016.

Addison, Paul. 'Churchill, Sir Winston Leonard Spencer (1874–1965)'. *Oxford Dictionary of National Biography*. Oxford University Press, 2004. Online edition, Jan 2011, available at http://www.oxforddnb.com/view/article/32413, accessed 8 June 2012.

Andersen, Casper. *British Engineers and Africa, 1875–1914*, Kindle edition. London: Pickering & Chatto, 2011.

Anthony, Frank. *Britain's Betrayal in India: The Story of the Anglo-Indian Community*. Bombay: Allied Publishers, 1969.

Arnold, David. '"An Ancient Race Outworn": Malaria and Race in Colonial India, 1860–1930'. In *Race, Science and Medicine, 1700–1960*, edited by Waltraud Ernst and Bernard Harris, pp. 123–43. London and New York: Routledge, 1999.

———. *Science, Technology and Medicine in Colonial India*. The New Cambridge History of India, III.5. Cambridge: Cambridge University Press, 2000.

———. *Everyday Technology: Machines and the Making of India's Modernity*. Chicago and London: University of Chicago Press, 2013.

Bagal, Jogesh Chandra. *Pramatha Nath Bose*. New Delhi: Sushama Sen on behalf of P. N. Bose Centenary Committee, 1955.

Bahl, Vinay. *The Making of the Indian Working Class: A Case of the Tata Iron and Steel Company, 1880–1946*. New Delhi, Thousand Oaks, and London: Sage Publications, 1995.

Bassett, Ross. 'MIT-Trained Swadeshis: MIT and Indian Nationalism, 1880–1947'. *Osiris*, vol. 24, no. 1 (2009): 212–30.

———. *The Technological Indian*, Kindle edition. Cambridge, MA, and London: Harvard University Press, 2016.

Basu, Aparna. 'Technical Education in India, 1854–1921'. In *Essays in the History of Indian Education*, pp. 39–59. New Delhi: Concept, 1982.

Beaglehole, T. H. 'From Rulers to Servants: The I.C.S. and the British Demission of Power in India'. *Modern Asian Studies*, vol. 11, no. 2 (1977): 237–55.

Bear, Laura Gbah. 'Miscegenations of Modernity: Constructing European Respectability and Race in the Indian Railway Colony, 1857–1931'. *Women's History Review*, vol. 3, no. 4 (1994): 531–48.

Bear, Laura. *Lines of the Nation: Indian Railway Workers, Bureaucracy, and the Intimate Historical Self.* New York: Columbia University Press, 2007.

Black, John. 'The Military Influence on Engineering Education in Britain and India, 1848–1906'. *The Indian Economic and Social History Review*, vol. 46, no. 2 (2009): 211–39.

Brassley, Paul. 'The Professionalisation of English Agriculture?'. *Rural History*, vol. 16, no. 2 (2005): 235–51.

Buchanan, R. A. 'Institutional Proliferation in the British Engineering Profession, 1847–1914'. *The Economic History Review*, New Series, vol. 38, no. 1 (February 1985): 42–60.

———. *The Engineers: A History of the Engineering Profession in Britain, 1750–1914.* London: Jessica Kingsley, 1989.

Buettner, Elizabeth. *Empire Families: Britons and Late Imperial India.* Oxford: Oxford University Press, 2004.

Cameron, J. G. P. *A Short History of the Royal Indian Engineering College: Coopers Hill.* N.p.: Coopers Hill Society, private circulation, 1960.

Campion, David A. 'Railway Policing and Security in Colonial India, c. 1860–1930'. In *Our Indian Railway: Themes in India's Railway History*, edited by Roopa Srinivasan, Manish Tiwari, and Sandeep Silas, pp. 121–53. New Delhi: Foundation, 2006.

Caru, Vanessa. 'The Creation of a "Corps d'État"? Indian and European Engineers of the Bombay Public Works Department (1860's–1940's)'. Paper presented at the midterm workshop of the ENGIND project under the French National Research Agency, New Delhi, 11–12 January 2016.

Chakrabarti, Pratik. *Western Science in Modern India: Metropolitan Methods, Colonial Practices.* Delhi: Permanent Black, 2004.

———. '"Signs of the Times": Medicine and Nationhood in British India'. *Osiris*, vol. 24, no. 1 (2009): 188–211.

Chandavarkar, Rajnarayan. 'Industrialization in India before 1947: Conventional Approaches and Alternative Perspectives'. *Modern Asian Studies*, vol. 19, no. 3 (Special Issue, 1985): 623–68.

Chandavarkar, Rajnarayan. *The Origins of Industrial Capitalism in India: Business Strategies and the Working Classes in Bombay, 1900–1940*. New Delhi: Foundation Books (in arrangement with Cambridge University Press), 1994.

Chaudhuri, B. N., M. I. E. 'A Short History of the Growth and Development of the Institution of Engineers (India)'. In *Demicenturion*, IEI Commemorative volume [1969], pp. 43ff. Volume consulted at the Library of the IEI, at its Kolkata Headquarters at 8 Gokhale Road.

Chibber, Vivek. *Locked in Place: State-Building and Late Industrialization in India*. New Delhi: Tulika Books, 2004.

Dalrymple, William. *Nine Lives: In Search of the Sacred in Modern India*. London, Berlin, and New York: Bloomsbury, 2009.

Derbyshire, Ian. 'The Building of India's Railways: The Application of Western Technology in the Colonial Periphery 1850–1920'. In *Technology and the Raj: Western Technology and Technical Transfers to India 1700–1947*, edited by Roy MacLeod and Deepak Kumar, pp. 177–215. New Delhi, Thousand Oaks and London: Sage Publications, 1995.

Dewey, C. J. 'The Education of a Ruling Caste: The Indian Civil Service in the Era of Competitive Examination'. *The English Historical Review*, vol. 88, no. 347 (April 1973): 262–85.

Downey, Gary Lee and Juan C. Lucena. 'Knowledge and Professional Identity in Engineering: Code-Switching and the Metrics of Progress'. *History and Technology*, vol. 20, no. 4 (December 2004): 393–420.

Edgerton, David. *The Shock of the Old: Technology and Global History since 1900*. London: Profile, 2008 [2006].

———. 'Innovation, Technology or History: What Is the Historiography of Technology About?'. *Technology and Culture*, vol. 51, no. 3 (July 2010): 680–97.

Elwin, Verrier. *The Story of Tata Steel*. Bombay, n.p.: 1958. British Library Shelfmark W 3092.

Ewing, Ann. 'The Indian Civil Service 1919–1924: Service Discontent and the Response in London and in Delhi'. *Modern Asian Studies*, Vol. 18, no. 1 (1984): 33–53.

Ghosh, Parimal. *Colonialism, Class and a History of the Calcutta Jute Millhands 1880–1930*. Hyderabad: Orient Longman, 2000.

Gilmartin, David. 'Scientific Empire and Imperial Science: Colonialism and Irrigation Technology in the Indus Basin'. *The Journal of Asian Studies*, vol. 53, no. 4 (November 1994): 1127–49.

Gispen, C. W. R. 'German Engineers and American Social Theory: Historical Perspectives on Professionalization'. *Comparative Studies in Society and History*, vol. 30, no. 3 (July 1988): 550–74.

Gispen, Kees. *New Profession, Old Order: Engineers and German society, 1815–1914*. Cambridge: Cambridge University Press, 1989.

Goldstein, Jan. 'Foucault among the Sociologists: The "Disciplines" and the History of the Professions'. *History and Theory*, vol. 23, no. 2 (May 1984): 170–92.

Goswami, Manu. *Producing India: From Colonial Economy to National Space*. Chicago and London: University of Chicago Press, 2004.

Habib, S. Irfan and Dhruv Raina, eds. *Social History of Science in Colonial India*. New Delhi: Oxford University Press, 2007.

Harris, F. R. (with Lovat Fraser). *Jamsetji Nusserwanji Tata: A Chronicle of his Life*, 2nd edition. Bombay: Blackie & Son, 1958.

Harrison, Mark. *Public Health in British India: Anglo-Indian Preventive Medicine 1859–1914*. Cambridge, New York, and Melbourne: Cambridge University Press, 1994.

———. 'Science and the British Empire'. *Isis*, vol. 96, no. 1 (March 2005): 56–63.

Headrick, Daniel R. 'The Tools of Imperialism: Technology and the Expansion of European Colonial Empires in the Nineteenth Century'. *Journal of Modern History*, vol. 51, no. 2 (June, 1979): 231–63.

———. *The Tentacles of Progress: Technology Transfer in the Age of Imperialism, 1850–1940*. New York and Oxford: Oxford University Press, 1988.

Henry, Odile and Mathieu Ferry. 'Sociology of UG Students at IIT Kanpur: Education-Path and Placement according to Admission Status'. Presentation at the mid-term workshop of the ENGIND project under the French National Research Agency, New Delhi, 11–12 January 2016.

Huddleston, G. *History of the East Indian Railway*. Calcutta: Thacker, Spink and Co, 1906.

Hurst, H. E. 'MacDonald, Sir Murdoch (1866–1957)'. *Rev.* Elizabeth Baigent, *Oxford Dictionary of National Biography*. Oxford University Press, 2004. Available at http://www.oxforddnb.com/view/article/34709, accessed 8 June 2012.

Iftikhar-ul-Awwal, A. Z. M. *The Industrial Development of Bengal: 1900–1939*. Delhi: Vikas Publishing House, 1982.

Jeffery, Roger. 'Allopathic Medicine in India: A Case of Deprofessionalisation?'. *Economic and Political Weekly*, vol. 13, no. 3 (21 January 1978): 101–3, 105–13.

———. 'Recognizing India's Doctors: The Institutionalization of Medical Dependency, 1918–39'. *Modern Asian Studies*, vol. 13, no. 2 (1979): 301–26.

Joshi, Sanjay, ed. *The Middle Class in Colonial India*. New Delhi: Oxford University Press, 2010.

Kerr, Ian J. *Building the Railways of the Raj: 1850–1900*. Delhi: Oxford University Press, 1995.

———, ed. *Railways in Modern India*. New Delhi: Oxford University Press, 2001.

———. *Engines of Change: The Railroads That Made India*. Westport, CT and London: Praeger, 2007.

———, ed. *27 Down: New Departures in Indian Railway Studies*. New Delhi: Orient Longman, 2007.

Kipling, Rudyard. 'William the Conqueror'. In *The Second Penguin Book of English Short Stories*, edited by Christopher Dolley, pp. 61–94. London: Penguin, 2011 [1972].

Klingensmith, Daniel. *'One Valley and a Thousand': Dams, Nationalism, and Development*. New Delhi: Oxford University Press, 2007.

Kumar, Arun. 'Colonial Requirements and Engineering Education: The Public Works Department, 1847–1947'. In *Technology and the Raj: Western Technology and Technical Transfers to India 1700–1947*, edited by Roy MacLeod and Deepak Kumar, pp. 216–32. New Delhi, Thousand Oaks, and London: Sage Publications, 1995.

Kumar, Deepak. 'Racial Discrimination and Science in Nineteenth-Century India'. *Indian Economic and Social History Review*, vol. 19 (1982): 63–82.

———. *Science and the Raj, 1857–1905*. Bombay and Oxford: Oxford University Press, 1995.

———. 'Reconstructing India: Disunity in the Science and Technology for Development Discourse, 1900–1947'. *Osiris*, 2nd series, vol. 15 (2000): 241–57.

Lala, R. M. *The Creation of Wealth: The Tata Story*, paperback edition. Bombay: IBH, 1981.

———. *For the Love of India: The Life and Times of Jamsetji Tata*. New Delhi: Penguin/Portfolio, 2006 [2004].

———. *The Romance of Tata Steel*. New Delhi: Penguin/Viking, 2007.

Lardinois, Roland. 'Coaching the Masses, Creaming the Elites: The Education Market at Kota (Rajasthan)'. Paper presentation at the mid-term workshop of the ENGIND project under the French National Research Agency, New Delhi, 11–12 January 2016.

Layton, Edwin T., Jr. *The Revolt of the Engineers: Social Responsibility and the American Engineering Profession*. Baltimore: Johns Hopkins University Press, 1986.

Lourdusamy, John Bosco. 'College of Engineering, Guindy, 1794–1947'. In *Science and Modern India: An Institutional History, c. 1784–1947*, edited by Uma Das Gupta, vol. 15, part 4 of the series 'History of Science, Philosophy and Culture in Indian Civilization', chapter 15, pp. 429–50. Delhi: Pearson Longman, 2011.

Lowe, Rodney. *The Official History of the British Civil Service: Reforming the Civil Service*, vol. 1: *The Fulton Years, 1966–81*. Abingdon: Routledge, 2011.

MacLeod, Roy M. 'Holland, Sir Thomas Henry (1868–1947)'. *Oxford Dictionary of National Biography*. Oxford University Press, 2004. Available at http://www.oxforddnb.com/view/article/33945, accessed 11 June 2012.

MacLeod, Roy, and Deepak Kumar, eds. *Technology and the Raj: Western Technology and Technical Transfers to India 1700–1947*. New Delhi, Thousand Oaks, and London: Sage Publications, 1995.

Mahindra, K. C. *Rajendra Nath Mookerjee: A Personal Study*. Calcutta: Art Press, 1933.

Majumdar, Sumit K. *India's Late, Late Industrial Revolution: Democratizing Entrepreneurship*. Cambridge, UK, and New York: Cambridge University Press, 2012.

Meiksins, Peter. 'The "Revolt of the Engineers" Reconsidered'. *Technology and Culture* , vol. 29, no. 2 (April 1988): 219–46.

Metcalf, Barbara D., and Thomas R. Metcalf. *A Concise History of India*. Cambridge and New York: Cambridge University Press, 2003 [2002].

Millard, J. Rodney. *The Master Spirit of the Age: Canadian Engineers and the Politics of Professionalism*. Toronto: University of Toronto Press, 1988.

Misra, Maria. 'Colonial Officers and Gentlemen: The British Empire and the Globalization of "Tradition"'. *Journal of Global History*, vol. 3, no. 2 (July 2008): 135–61.

Mital, K. V. *History of the Thomason College of Engineering (1847–1949): On Which Is Founded the University of Roorkee*. Roorkee: University of Roorkee, 1986.

Moore, R. J. 'The Problem of Freedom with Unity: London's India Policy, 1917–47'. In *Congress and the Raj: Facets of the Indian Struggle 1917–47*, edited by D. A. Low, 2nd edition, pp. 375–403. New Delhi: Oxford University Press, 2004.

Muldoon, Andrew. *Empire, Politics and the Creation of the 1935 India Act: Last Act of the Raj*. Abingdon, Oxon: Ashgate, 2009.

Muthiah, S. 'Engineers Who Made History'. *The Hindu, Metro Plus*, 9 June 2008, online version. Available at http://www.thehindu.com/todays-paper/tp-features/tp-metroplus/engineers-who-made-history/article1418243.ece, accessed 3 September 2015.

Nandy, H., ed. *IEI Marches On*. Kolkata: Cdr. A. K. Poothia for Institution of Engineers (India), 2002 [1996].

Narayanan, N. C., Poonam Argade, and D. Parthasarathy. '"Dis(coursing) Water": Dams and Changing Engineering Paradigms in Western India'. Contribution to the mid-term workshop of the ENGIND project under the French National Research Agency, New Delhi, 11–12 January 2016.

Natesan, L. A. *State Management & Control of Railways in India: A Study of Railway Finance Rates and Policy during 1920–37.* Calcutta: University of Calcutta, 1946.

Nitesh, Ravi. 'The legacy of Sir Ganga Ram'. *Daily Times* (Pakistan) online, 17 April 2014. Available at http://www.dailytimes.com.pk/opinion/17-Apr-2014/the-legacy-of-sir-ganga-ram, accessed 3 September 2015.

Noble, David F. *America by Design: Science, Technology, and the Rise of Corporate Capitalism.* Oxford: Oxford University Press, 1979.

Nomura, Chikayoshi. 'Selling Steel in the 1920s: TISCO in a Period of Transition'. *Indian Economic and Social History Review*, vol. 48, no. 1 (2011): 83–116.

Oba, Hiruyoki, and Hrushikesh Panda, eds. *Industrial Development and Technology Absorption in the Indian Steel Industry: Study of TISCO with Reference to Yawata—A Steel Plant of Nippon Steel Corporation in Japan.* Mumbai: Allied Publishers, 2005.

Patel, Shirish B. 'Why Flyovers Will Fall: Decline of the Civil Engineering Profession in India'. *Economic and Political Weekly*, vol. 51, no. 20 (14 May 2016): 32–6.

Phalkey, Jahnavi. *Atomic State: Big Science in Twentieth-Century India.* Ranikhet: Permanent Black, 2013.

———. 'Introduction' ('Focus: Science, History, and Modern India'). *Isis*, vol. 104, no. 2 (June 2013): 330–6.

Potter, David C. 'Manpower Shortage and the End of Colonialism: The Case of the Indian Civil Service'. *Modern Asian Studies*, vol. 7, no. 1 (1973): 47–73.

———. *India's Political Administrators 1919–1983.* Oxford: Clarendon, 1986.

Prakash, Gyan. *Another Reason: Science and the Imagination of Modern India.* Princeton: Princeton University Press, 1999.

Prasad, Amit. *Imperial Technoscience: Transnational Histories of MRI in the United States, Britain, and India*, Kindle edition. Cambridge, MA: MIT Press, 2014.

Raina, Dhruv. *Visvesvaraya as Engineer-Sociologist and the Evolution of His Techno-Economic Vision.* Bangalore: National Institute of Advanced Studies, 2001.

———. *Images and Contexts: The Historiography of Science and Modernity in India.* Delhi and Oxford: Oxford University Press, 2003.

Ramnath, Aparajith. 'Industrial Experts in the Age of Indianisation: The European Engineering Firms of Calcutta, 1914–47'. Paper presentation at the mid-term workshop of the ENGIND project under the French National Research Agency, New Delhi, 11–12 January 2016.

———. 'International Networks and Aircraft Manufacture in Late-Colonial India: Hindustan Aircraft Limited, 1940–47'. IIM Kozhikode working paper no. IIMK/WPS/205/HLA/2016/17, July 2016.

Ray, Rajat K. *Industrialization in India: Growth and Conflict in the Private Corporate Sector 1914–47*. New Delhi: Oxford University Press, 1982 [1979].

Riddick, John F. *The History of British India: A Chronology*. Westport, CT: Praeger, 2006.

Rothermund, Dietmar. *An Economic History of India: From Pre-Colonial Times to 1991*, 2nd edition. London: Routledge, 1993.

Roy, Tirthankar. *The Economic History of India 1857–1947*, 2nd edn. New Delhi: Oxford University Press, 2006.

Sahni, Jogendra Nath. *Indian Railways: One Hundred Years, 1853 to 1953*. New Delhi: Ministry of Railways (Railway Board), 1953.

Sharma, Dinesh C. *Nehru: The Unlikely Hero of India's Information Technology Revolution*. NMML Occasional Paper, Perspectives in Indian Development, New Series 8. New Delhi: Nehru Memorial Museum and Library, 2013.

Sharma, Dinesh C. *The Outsourcer: The Story of India's IT Revolution*. Cambridge: MIT Press, 2015.

Sharma, Malti. *Indianization of the Civil Services in British India (1858–1935)*. New Delhi: Manak Publications, 2001.

Shinn, Terry. 'From "Corps" to "Profession": The Emergence and Definition of Industrial Engineering in Modern France', in *The Organization of Science and Technology in France 1808–1914*, edited by Robert Fox and George Weisz, pp. 183–208. Cambridge: Cambridge University Press, 1980.

Shukla, J. D. *Indianisation of All-India Services and Its Impact on Administration*. New Delhi: Allied Publishers, 1982.

Sohoni, Milind. 'Engineering Teaching and Research in IITs and Its Impact on India'. *Current Science*, vol. 102, no. 11 (10 June 2012): 1510–15.

Stein, Burton. *A History of India*, 2nd edn, edited by David Arnold. Oxford: Wiley-Blackwell, 2010.

Subramanian, Ajantha. 'Making Merit: The Indian Institutes of Technology and the Social Life of Caste'. *Comparative Studies in Society and History*, vol. 57, no. 2 (2015): 291–322.

Sundaram, Chandar S. '"Treated with Scant Attention": The Imperial Cadet Corps, Indian Nobles, and Anglo-Indian Policy, 1897–1917'. *The Journal of Military History*, vol. 77 (January 2013): 41–70.

Tandon, Prakash. *Punjabi Century: 1857–1947*. London: Chatto and Windus; Toronto: Clarke, Irwin and Co, 1961.

Tetzlaff, Stefan. 'State Policy, Technical Cooperation and Manpower Requirements: The Case of Indian Automotive Engineering around the Mid-20th century (c. 1947–1970)'. Paper presentation at the mid-term workshop of the ENGIND project under the French National Research Agency, New Delhi, 11–12 January 2016.

Tignor, Robert L. 'The "Indianization" of the Egyptian Administration under British Rule'. *The American Historical Review*, vol. 68, no. 3 (April 1963): 636–61.

Tomlinson, B. R. 'Colonial Firms and the Decline of Colonialism in Eastern India 1914–47'. *Modern Asian Studies*, vol. 15, no. 3 (1981): 455–86.

———. *The Economy of Modern India, 1860–1970*, The New Cambridge History of India, III.3. Cambridge: Cambridge University Press, 1996 [1993].

Upadhya, Carol. *Reengineering India: Work, Capital, and Class in an Offshore Economy*. Delhi: Oxford University Press, 2016.

Visvanathan, Shiv. *Organizing for Science: The Making of an Industrial Research Laboratory*. Delhi: Oxford University Press, 1985.

Watson, Garth. *The Civils: The Story of the Institution of Civil Engineers*. London: Thomas Telford, 1988.

Theses/Dissertations

Cuddy, Brendan P. 'The Royal Indian Engineering College, Cooper's Hill, (1871–1906): A Case Study of State Involvement in Professional Civil Engineering Education'. PhD thesis, London University, 1980.

Ramnath, Aparajith. 'Breaking Free: Technical Education Policy in India Immediately before and after Independence'. MSc dissertation, University of Oxford, 2007.

———. 'Engineers in India: Industrialisation, Indianisation and the State'. PhD thesis, Imperial College London, 2012.

Sebaly, Kim Patrick. 'The Assistance of Four Nations in the Establishment of the Indian Institutes of Technology, 1945–1970'. PhD thesis, University of Michigan, 1972.

Srinivasa Rao, Y. 'Electrification of Madras Presidency, 1900–1947'. PhD thesis, Indian Institute of Technology Madras, 2007.

WEBSITES

'David Sassoon Library and Reading Room: History'. Available at http://www.davidsassoonlibrary.com/history.html, accessed 2 July 2012.

'ENGIND: Engineers and Society in Colonial and Post-Colonial India', website of international project under the French National Research Agency. Available at http://engind.hypotheses.org/, accessed 15 October 2016.

'Engineering Heritage: Past Presidents'. Available at http://heritage.imeche. org/Biographies/pastpresidents, accessed 14 April 2012.

'History of C. P. W. D.', at 'Organization', 'Historical Background'. Available at http://cpwd.gov.in, accessed 21 July 2012.

'History: Pakistan Engineering Congress an Overview of 94 years'. Available at http://pecongress.org.pk/history.php, accessed 3 July 2012.

'IMA in Retrospect'. Available at http://www.ima-india.org/IMA_history. html, accessed 12 April 2012.

'Indian Railways: Indian Railway Conference Association: History'. Available at http://www.indianrailways.gov.in/railwayboard/uploads/directorate/ IRCA/index.jsp, accessed 17 July 2012.

'Sir Bezonji Mehta (1840–1927)'. Available at http://www.tatacentralarchives. com/history/biographies/02%20bezonjimehta.htm, accessed 1 July 2012.

'Sir Lakshmi Pati Misra', in 'Luminaries'. Available at http://www.iitr.ac.in/ institute/pages/Heritage+Luminaries.html, accessed 22 July 2011.

'Vanguards'. Available at http://www.tatasteel100.com/people/vanguards. asp, accessed 1 July 2012.

Fenwick, S. C., 'Corps History – Part 10: Indian Sappers (1740–1947)'. Website of the Royal Engineers Museum, Kent, available at http://www.remu-seum.org.uk/corpshistory/rem_corps_part10.htm, last accessed 4 June 2008. The source cited for the article by the website is G. Napier, *Follow the Sapper: An Illustrated History of the Corps of Royal Engineers* (Chatham: Institution of Royal Engineers, 2005).

Indian Railways Fan Club, 'Books, Timetables, Videos etc. – II'. Available at http://www.irfca.org/faq/faq-books2.html, accessed 3 September 2015.

Institute of Technology, Banaras Hindu University. 'Heritage'. Available at http://www.itbhu.ac.in/itbhu/heritage.shtml, accessed 9 June 2012.

Das, Lala Ram, 19
Department of Commerce and
    Industry, India, 50
deputation, of engineers, 119
Derbyshire, Ian, 57n3, 63
division of labour, 192
division of services, 38
Domiciled Europeans, 124, 166, 171;
    status of, 36–7
Dominion Iron and Steel Company
    (Canada), 188
Donner, Patrick, 129, 131–2
Dorman, W. S., 66, 67, 82
Doyle, Pat., 64
dual loyalties, of engineers, 21
Dutch East Indies, 11
duties of engineers, 118
dyarchy, principle of, 43–6, 112, 208
Dyer, General, 44
Dyer, G. H. Thiselton, 83

East India (Constitutional Reforms).
    See *Montagu–Chelmsford Report*
    (1918)
East India Company, 27–8
East India Railway (EIR), 64, 121,
    153, 156, 166; Special Class
    Apprentice (SCA) programme,
    175–6
economic nationalism, 53, 91, 93,
    231
Edgerton, David, 8
education system, in India:
    centralized institutes, idea of,
    9; competitive examinations,
    in India, 39; engineering
    education, 6; 'European
    schools' curriculum and, 37;
    IEI's Associate Membership
    examinations, 78–9; medical
    education, 9; of professional

engineers, 234; technical
    education facilities, 11
efficiency, of engineers, 72, 76, 84,
    95, 129, 131, 155, 160–1, 180,
    227–8, 232
electrical engineering, 6, 203, 220
employment: in large-scale
    industry, 53; major employers,
    of engineers in India, 59; of non-
    Europeans, 38, 45
Empress Mills (textiles), Nagpur,
    195
ENGIND (Engineers and Society in
    Colonial and Postcolonial India)
    project, 128n110
engineering education, in India,
    6; IEI's Associate Membership
    examinations, 78–9; IEI's role in,
    76; qualification for individuals,
    78
engineering profession, 1–2,
    57; after World War I, 69–76;
    British composition of, 59–63;
    Chartered Engineer (India), 80;
    development during Nehruvian
    era, 6; economic policies,
    impact of, 9; emergence of
    an Indian identity in, 23, 77;
    expatriate engineers, 58–64;
    government policies on, 69–76;
    IEI and growth of, 76–81; Indian
    institution, formation of, 69–76;
    Indianization of, 6; informal
    licensing function, 60; interwar
    experience, 7–8; *laissez faire*
    policy, 71; local organization,
    64–9; major employers,
    of engineers in India, 59;
    metropolitan institutions, 58–64;
    qualities required for working in
    India, 116

Indian Civil Service (ICS), 12–13, 68, 97, 115; competitive examination system used for, 102; European-dominated nature of, 13; Indianization of, 13, 30; pay and emoluments, 38, 40; recruitment and training, 38–9; reform in recruitment in, 37; reforms in composition of, 30
Indian Companies Act (1935), 76
Indian Councils Act (1909), 29
Indian Educational Service, 14
Indian Engineering, 1, 22, 64, 79, 83, 124
Indian engineering colleges, 34, 96, 103, 109, 112, 124, 142, 145, 177; opportunities for the graduates of, 110–11
Indian Engineering Society, 82
Indian engineers: America-trained, 11; British attitudes to, 126; in interwar India, 16–21; status of, 123–33
Indian Fiscal Commission (1921–2), 51
Indian Industrial Commission (IIC), 50, 71, 77, 162, 204; recommendations by, 51; vision of, 51
Indian Institutes of Technology (IITs), 233–4; 'coaching' industry, 234; Joint Entrance Exam (JEE), 234; origins of, 234–5
Indian institution, formation of, 69–76; movement for, 72
Indian Institution of Engineers, 72–3
Indianization, of public services: 1919 reforms and, 42–7; British officers and, 44; bureaucratic set-up of, 31;

constitutional reforms and, 30–49; engineering profession, 5; expanded legislatures and, 37–42; factors influencing, 54–5; government administration, 32; Government of India Act (1935), 47–9; government's attitude to, 27; history of, 27–30; implementation of, 45; industrialization and, 12–16; Institution of Engineers India (IEI) and, 84–92; Islington Commission's approach to, 37–42, 110–11; Lee Commission (1923–4) on, 42–7; medical profession, 9–10; mutual mistrust on, 159–65; nationalist movement, rise of, 13; politics of, 42–7; Public Works Department (PWD), 105, 108–10, 135; of railways, 137, 157–65; Tata Iron and Steel Company (TISCO), 184–5
Indian Medical Association, 75
Indian Medical Service (IMS), 48, 75, 98; Indianization of, 9
Indian 'modernity,' creation of, 3
Indian National Congress, 28–9, 42; contacts with Indian businessmen, 53; demand for reform in ICS recruitment, 37; National Planning Committee of, 91
Indian-owned companies, importance of, 53
Indian Police, 13, 33, 35, 39–40, 46, 48
Indian Railway Gazette, 22, 155–6, 161, 175
Indian Railways Conference Association (IRCA), 178

Indian Railway Service of Engineers
(IRSE), 143
Indian science and technology:
during colonial period, 3–4, 3n8,
7; colonial technology and, 8;
developments in, 5; history of,
8–12; interwar period, 7
Indian Science Congress, 75–6
Indian scientific community, growth
of, 75
Indian Service of Engineers (ISE),
13, 108–9, 125–6, 133, 226; direct
recruitment and promotion in,
111; *India Office List*, 113–14,
178; Irrigation branch of,
113; lists of officers of, 96n3;
percentage of Indians in, 112;
rankings in, 109–10; recruitment
by the Secretary of State, 113;
Roads and Buildings branch of,
113; routes into, 111
Indian Tariff Board (ITB), 52, 203, 212
Indian 'techie,' phenomenon of, 224
India Office List, 113–14, 147, 178
industrial banks, creation of, 50–1
industrial education, 50, 184, 197,
232, 234
industrial engineering, 6, 90, 221
industrial engineers, 90; profession
of, 21
industrial enterprises, 4, 14, 16, 54,
69, 182, 184, 220, 224, 228, 231,
235
industrialization, in India, 4; capital
investment for, 53; and financial
assistance to industries, 51;
government's industrial policy,
evolution of, 51; Indianization
of, 12–16; industries granted
protection, categories of,
52; Institution of Engineers

India (IEI), 84–92; interwar
policies, 49–54, 232; large-scale
industry, growth of, 49–54;
limits under colonial rule, 15;
ownership pattern of industries,
52; protective tariffs, against
imports, 51–2
industrial production, 49
industries granted protection,
categories of, 52
Innes, Charles, 164–5
Institution of Civil Engineers (ICE),
60, 65, 71, 82, 93; Associate
Membership of, 63, 103;
membership trends in, 70
Institution of Electrical Engineers
(IEE), 22n69, 60, 65;
membership trends in, 71
Institution of Engineers India (IEI),
58, 71, 75, 178, 225; Associate
Membership examinations, 78–9;
body corporate of, 78; educational
qualifications, 78; establishment
of, 76; formation of, 84, 93;
functions of, 76; goals of, 76; and
growth of Indian engineering
profession, 76–81; Indianization
and industrialization, 84–92;
*Journal of the Institution of
Engineers (India)*, 77; objectives
of, 76–7; Presidents of, 86–7;
primary source of legitimacy, 76;
Professional Conduct Rules, 80;
recommendations of, 77; role
in engineering education, 76;
system of graded membership,
78; titles of some papers
discussed at meetings of, 89
Institution of Mechanical
Engineers (IMechE), 60, 62, 65;
membership trends in, 70

multinational experts, 185–200;
organization structure of, 190–1;
plea for economic protection,
203; registration of, 185, 187;
steel-making, culture of, 185–
200; strategies in recruiting and
training, 183; technical experts
of, 183–4; training facilities,
203; workforce of, 192; *see also*
Jamshedpur Technical Institute
(JTI)
Tata, Jamsetji Nusserwanji, 185, 187
technical education, 11, 234
technical expertise, 2, 6, 134
technical practitioners, 2, 5, 7, 18,
25, 57, 232; Indianization of, 222
techno-bureaucrats, 146
technological identity, 10
technology transfer, in colonial
India, 183
Thomason College of Engineering,
Roorkee, 101–2, 110, 124
Tilak, Bal Gangadhar, 42
Tomlinson, B. R., 53n101
traditional knowledge, 3
'transferred' fields, 43

Turnour, Edward, 132
Tutwiler, T. W., 193–4, 197, 199, 222

unskilled labour, 192n28, 201
Upper Subordinates, in railways,
171; percentage shares among,
171n107

Varma, B. P., 83
Victoria, Queen, 28
Visvanathan, Shiv, 51
Visvesvaraya, M., 20n61, 125
vocation, of engineering, 237
voluntary retirement, policy of, 171

Wakefield, George, 126
Weld, C. M., 186–7
Wells, R. G., 187, 193, 196
'Western' technologies, 11
World War I, 14, 15, 23, 33n21, 34,
36, 42, 49, 53, 62n22, 69, 70,
93, 105, 112, 133, 136, 143, 152,
171n107, 183, 189, 200, 201, 225,
231
World War II, 11, 24, 91, 98, 168,
173, 174, 180, 220, 228, 236

Aparajith Ramnath is a historian of science, technology, and industry in South Asia. He is Assistant Professor (Humanities and Liberal Arts) at the Indian Institute of Management Kozhikode, where he teaches courses on Indian society, business history, and the global history of industrialization.